I0045913

PROSPECTOR'S
FIELD-BOOK AND GUIDE

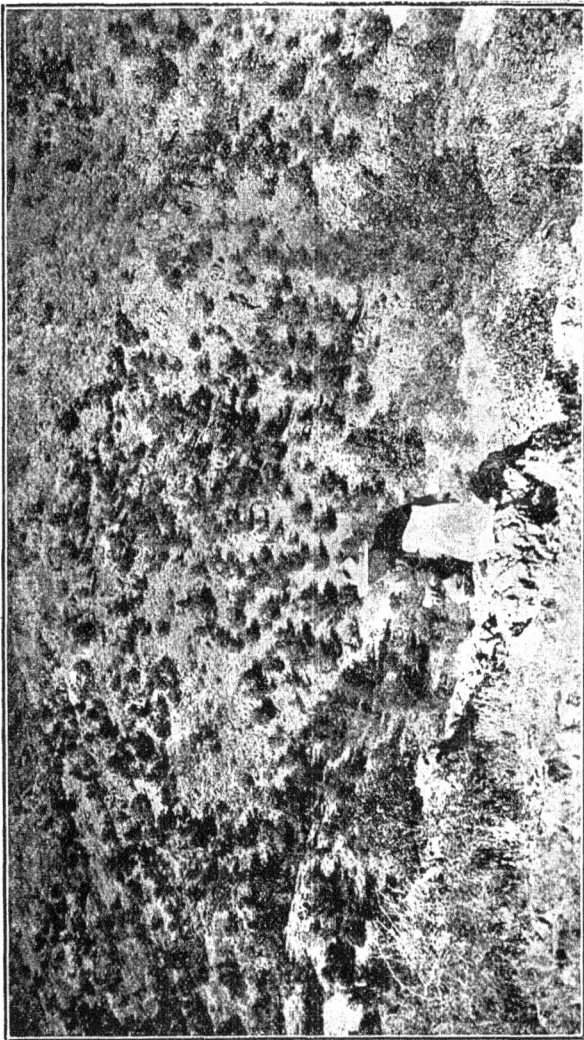

AN OUTCROP: THE SURFACE INDICATION OF AN OREBODY

PROSPECTOR'S
FIELD-BOOK AND GUIDE

IN THE SEARCH FOR AND THE EASY DETERMINA-
TION OF ORES AND OTHER USEFUL MINERALS

BY

H. S. OSBORN

Author of "Metallurgy of Iron and Steel," "Practical
Manual of Minerals, Mines and Mining"

NINTH EDITION
THOROUGHLY REVISED AND ENLARGED

BY

M. W. von BERNEWITZ

Metallurgist and Technical Journalist; Formerly Assistant Editor
of "Mining and Scientific Press" and "The Mines
Handbook;" Contributor to "The Mineral
Industry" and other Publications

ILLUSTRATED BY FIFTY-SEVEN ENGRAVINGS

NEW YORK
HENRY CAREY BAIRD & CO., Inc.
Publishers of Mechanical and Industrial Books
2 WEST 45TH STREET
1920

COPYRIGHT BY
HENRY CAREY BAIRD & CO.
1910

COPYRIGHT BY
HENRY CAREY BAIRD & CO., INC.
1920

PRINTED IN U. S. A.

PREFACE TO THE NINTH EDITION

Another edition of this well-known book is justified. The search for minerals is a fairly steady business, affected at times by market conditions. The past five years was one of intense demand and exploration, during which time many of those in the field found that they had to seek ores with which they were unfamiliar, and no text-book was available that covered all of these subjects in concise manner. The eighth edition was published in 1910, since when there have been many changes in conditions, ideas, and methods, so the matter herein has been brought up to date. It is not claimed that all of the new material given is original, as a number of brief descriptions of ore deposits and tests have been abstracted from those valuable sources of information — Bulletins of the United States Geological Survey and Bureau of Mines, various State Mining Bureau and Schools of Mines publications, and the technical press, consisting of the ' Mining and Scientific Press ' and ' Engineering and Mining Journal '; also ' The Mines Handbook.' Under the chapter headings will be found a general re-arrangement, with many additions, such as lists of suitable outfits, new field tests, new illustrations, notes on sampling, an explanation of the unit system of selling ores, and an entirely new chapter on the alloy minerals; in fact the book is considerably changed and enlarged in its scope. As a guide and aid to prospecting, the principal characteristics of certain ore deposits in various parts of the world are briefly discussed; this in addition to general instructions for field work. During my five years of practical experience in New Zealand, twelve in Australia, and one in California, with six years as assistant editor of the ' Mining and Scientific Press ' at San Francisco, and one year with ' The Mines Handbook ' at New York, I was particularly observant of general conditions in many mining regions, and answered many hundreds of queries on minerals, metals, and methods. It is to be hoped that prospecting will continue to be an important accessory to mining, which is one of the basic industries of the United States and of the World, and that this ninth edition will be of some assistance to the profession.

M. W. von Bernewitz.

New York, January 10, 1920,

CONTENTS

PROSPECTOR'S FIELD-BOOK
AND GUIDE

INTRODUCTION

Much disappointment and financial loss has been experienced by those who have gone prospecting, through their lack of knowledge of the probable locality, mode of occurrence, and characteristics of the various commercial minerals, also of the best methods of testing them when found. Many great mines are the result of accidental discoveries, and often by men who were quite ignorant of the value of minerals. Many valuable deposits have been overlooked, or abandoned, due to a lack of knowledge of mineralogy and mining. It might be said right here that the search for minerals is not what it used to be. The prospector is an essential factor in the future of mining. The one that was full of optimism, grubstaked by a storekeeper, went out with a burro or two, and promised payment to an assayer when rich ore was struck, exists only in small numbers these days. He had large areas over which he could wander, played an important part in the industry, and discovered valuable orebodies. As mining districts were developed, his field became restricted, yet there are many regions to be explored, but scientifically. It has been said that prospectors are born, not made, and should have the instinct for exploration. To a certain extent this is true, but the prospector of today must have some technical training, so as to understand the fundamentals of the business in which he is engaged. Prospecting involves a great deal of personal hardship, and requires both a practical and theoretical knowledge, tenacity of purpose, and great foresight. Rule-of-thumb methods are no longer applicable to prospecting. This is proved by the search for the rarer minerals during the years 1914 to 1919, when a knowledge of the rocks enclosing such was absolutely necessary if deposits were to be found. It must be remembered that gold and silver are not the only metals worth finding, although gold is independent of market fluctuations, and silver somewhat so. It is for this reason that such institutions as the Mackay School of Mines at Reno, Nevada; the School of Mines at Moscow, Idaho; that at Olympia, Washington; and others each year have free summer sessions extending over several weeks, during which prospectors, miners, and others interested may learn the fundamentals of mineralogy, geology, use of explosives, assaying, and chemistry. When such a grounding has been grasped, those about to or who have already been in the field, will be far more qualified to prospect, develop, value, and dispose of their ore deposits. Some knowledge of mining laws should also be obtained from one of the several excellent little works on this subject. One of the troubles with prospectors in the past has been a lack of judgment in not better opening his prospect himself before offering it to capital, and his inability to value it like an engineer. It is no use the prospector or miner deprecating scientific methods of exploration; such has been and is often done. If he is able to buy them, he should now be equipped with an automobile or motor-cycle, a portable testing outfit, and the knowledge before

discussed. He should also secure copies of the excellent
Bulletins published free by the United States Geological
Survey and Bureau of Mines, those issued by the various
Western States departments and Schools of Mines (those
mentioned and the institutions in Arizona, Colorado, Montana, New Mexico, Oregon, South Dakota, and in Ontario), and subscribe to at least one of the technical
weeklies. As instances of the necessity of having a
knowledge of geology take the mining boom at Divide in
Nevada during 1919. Work was first done there 17 years
before, but little was found. Further prospecting in 1918
emphasized the fact that the silver orebodies were in a
breccia, to reach which the overlying rhyolite must be
passed through. The intense gold and copper development at Oatman and Jerome, Arizona, in 1915 and 1917,
necessitated a knowledge of geologic conditions (the
United Eastern mine at the former place was the result
of study of geology, and intelligent mine development) ;
also the oil boom in Texas during 1919, although prospecting for petroleum is more or less of a special nature.
Every sign of mineralization should be examined. A
railroad excavation resulted in discovery of the great
silver deposits at Cobalt, Ontario, although not until many
people had passed it by. As for an instance of how valuable minerals may be overlooked, even by skilled men,
take the case of the zinc carbonate ore of Leadville, Colorado. Although this center has been one of the most important mining areas in the country for the last 40 years
and during that time has produced oxidized ores of gold,
silver, and lead, as well as sulphide ores of these metals
and of zinc and copper, it was not until 1910 that oxidized
zinc ore was discovered there in large quantity. The

rarer minerals were mentioned above; by these are meant the non-ferrous group such as antimony, chromite, magnesite, manganese, molybdenite, pyrite, tungsten, and others. The rocks surrounding these are different in most cases, so should be known; while the prospector should keep in touch with the market demand. So finally it might be reiterated that the search for ore deposits is becoming a specialized profession, and they that keep this fact in mind are the ones most likely to benefit by it.

The following list of metals and minerals, prepared by C. K. Leith of the U. S. Geological Survey, gives an idea of the resources of America, and their future possibilities:

The United States is more nearly self-sustaining in regard to mineral commodities as a whole than any other country on the globe.

1. Minerals of which there is an adequate supply or exportable surplus in the United States:

A. Minerals of which the exportable surplus dominates the world situation:

      Copper.

      Petroleum.

B. Minerals of which the exportable surplus constitutes an important but not a dominant factor in world trade:

Sulphur.	Iron and steel.
Phosphates.	Coal.
Silver.	Cement.

      Uranium and radium.

C. Minerals of which the exportable surplus is not an important factor in world trade. Small amounts of these minerals have been and will doubtless

continue to be imported because of special grades, back haul, or cheaper sources of foreign supply, but these imports are for the most part incidental:
Lead.
Zinc.
Aluminum and bauxite.
Gold.
Tungsten.
Molybdenum.
Asphalt and bitumen.
Pyrite.
Barite.
Fluorspar.
Building stone (excepting Italian marble).
Cadmium.
Gypsum.
Lime.
Tripoli and diatomaceous earth.
Mineral paints (except umber, sienna, and ocher from France and Spain).
Pumice.
Garnet.
Salt (except special classes).
Talc.
Arsenic.
Bismuth.
Bromine.
Artificial abrasives, corundum, and emery (except naxos emery).
Fuller's earth.
Mercury.

2. Minerals for which the United States demand must continue to be met by imports:

A. Minerals for which the United States must depend almost entirely on other countries:

Tin.

Nickel.

Platinum and metals of the platinum group.

B. Minerals for which the United States will depend on foreign sources for a considerable fraction of the supply:

Antimony.	Asbestos.
Vanadium.	Ball clay and kaolin.
Zirconium.	Chalk.
Mica.	Cobalt.
Monazite.	Naxos emery.
Graphite.	Grinding pebbles.

3. Minerals normally imported into the United States that in future can be largely produced from domestic sources if it seems desirable:

A. The following minerals were mainly imported before 1914, but under war conditions the domestic resources have been developed to such an extent that the United States can become self-sustaining if desirable, though at so great a cost that a protective tariff will be necessary if these industries are to survive:

Nitrates (except potassium nitrates).

Potash.

Manganese.

Chromite.

Magnesite.

CHAPTER I

PREPARATORY INSTRUCTION

Any chemically homogeneous substance that is a constituent of the earth's crust is a mineral. When two or more minerals occur together and form large masses, they constitute rocks. Those which contain sufficient metallic minerals to be profitably mined are classed as ores. The properties of minerals are numerous. Some of these, such as form, bulk, hardness, color, etc., are readily perceived; others, such as chemical nature, crystalline structure, and behavior towards light and heat, are not so apparent, and can only be determined by systematic investigation. The most significant characteristics are chemical composition, crystalline form, and density.

The minerals that are the principal constituents of rocks are the following:

1. **Those containing silica:** such as quartz; the feldspars; the micas; hornblende; pyroxene; talc; serpentine; chlorite.

2. **Carbonates:** such as carbonate of lime or calcite; carbonate of lime and magnesia or dolomite.

3. **Sulphates:** such as sulphate of lime or gypsum.

The special features of these, and of other less common mineral constituents, may be learned from a textbook on mineralogy. Following are the prominent characteristics of the minerals commonly encountered by the prospector:

7

Quartz. Occurs in crystals; also massive with a glassy luster. It is too hard to be scratched with a knife. It varies in color from white or colorless to black, and in transparency, from transparent quartz to opaque. It has no cleavage; that is, it breaks as easily in any direction the same as glass.

There are many varieties of quartz, of which may be mentioned: **Limpid quartz,** clear and colorless; **amethyst,** violet crystals; **agate,** presenting various colors arranged in parallel bands, straight, curved, or zigzag; **chalcedony,** transparent or translucent, and varying in color from white to gray, blue, brown and other shades; **flint,** massive, dark and dull color, edges translucent; **hornstone,** resembles flint, but differs from it in being more brittle, in breaking with a splintery, uneven fracture, and in not being so hard as quartz; **basanite, Lydian stone, or touchstone,** velvety black, more opaque than hornstone used for testing the purity of gold.

Opal is also a form of silica. Its special features are discussed on page 307. It can be scratched by quartz, and thus be distinguished from the latter. Opal is frequently met with in seams of certain volcanic rocks. Sometimes it occurs in limestone and also in metalliferous veins. The specific gravity and hardness of opal are slightly less than that of quartz.

Feldspar. The feldspars have a luster nearly like quartz, but are often somewhat pearly on smooth faces, are nearly as hard as quartz, with about the same specific gravity (2.4 to 2.6); and in general have light colors, mostly white or flesh-colored, though occasionally dark-grey, brownish, or green. They differ from quartz in having a perfect cleavage in one direction, yielding under the hammer a smooth lustrous surface and another nearly as perfect in a second direction inclined 84° to 90° to the first; in being fusible before the blowpipe, though not easily so; also in composition, the feldspars consisting of silica combined with alumina and an alkali — this alkali being either potash, soda, or lime, or two or all of them combined. Included in this group are a number of distinct kinds or species. These species differ in the proportion of silica (the acid) to the other ingredients (bases), and in the particular alkali (potash, soda, or lime) predominating.

The most important kinds are:

Orthoclase, or common feldspar, a **potash** feldspar. The cleavages make a right angle with one another, whence the name, signifying cleavage at a right angle.

In the following kinds, the cleavage forms angles of 84° to 87°; hence they are sometimes called **anorthic** feldspars or **plagioclastic** feldspars.

Albite, a **soda** feldspar, colorless and transparent, or translucent, and various shades of red, yellow, green and gray.

Oligoclase, a soda-lime feldspar, the soda predominating. Color — generally whitish or grayish, with shades of green and yellow.

Labradorite, a lime-soda feldspar, often iridescent. Color — usually ash or greenish-gray, but frequently various shades of green, yellow, and red. Sometimes the smaller crystals are colorless.

Anorthite, a lime feldspar, transparent and colorless, or translucent and grayish or reddish.

Feldspars are essentially constituents of volcanic and crystalline igneous rocks, orthoclase being typical of granite, syenite, gneiss, and trachyte, usually in association with quartz. Labradorite is the feldspar of basalts and dolerites in microscopic crystals. It also forms enormous rockmasses in Labrador. Oligoclase may be associated with orthoclase in granite, and is the feldspathic constituent of **diorite** and **diabase**. **Andesite** is the feldspar of the trachytes of the Andes, South America. **Albite** is chiefly found in crystalline schists and also in granite veins. **Anorthite** is best developed in the crystalline limestone rocks of Vesuvius, and also occurs in some basalts.

Micas. This embraces a group of minerals whose most marked feature is a highly laminated structure, and admit of being split into leaves thinner than paper. They are colorless to brown, green, reddish, and black, and occur either in small scales disseminated throughout rocks — as in granite — or in large plates. The micas are silicates of alumina with either potash, magnesia or iron, and some other ingredients.

The most important species of mica are:

Muscovite. This is the common mica, which, in the form of clear or slightly smoky colored plates, is used in the doors of stoves and lanterns. In Russia it was used for the windows of houses and this gave the name to the mineral of ' Muscovy glass,' whence the mineralogical name of muscovite.

Muscovite is a potash mica usually occurring in rhombic or six-sided tabular crystals. In many rocks the crystals are but poorly developed, or only represented by irregularly shaped scales; cleavage is basal and very perfect; color mostly silvery-white, seldom, but occasionally, darkbrown or black. Before the blowpipe it whitens and fuses on thin edges to a gray or yellow glass. Muscovite is not decomposed by sulphuric or hydrochloric acid.

Phlogopite, a magnesia mica of light brown, or copper-red, and sometimes white, color. It is common in limestone or in serpentine rocks and in dolomites.

Biotite. This includes most of the magnesia-iron micas. Color — black or dark green. .Very thin laminæ appear brown, greenish, or red by transmitted light. Luster, pearly; hardness, 2.5 to 3; specific gravity, 2.7 to 3.1. The basal cleavage is highly perfect, and the laminæ are flexible and elastic as in other members of the mica group. It is only slightly acted upon by hydrochloric acid, but is decomposed by sulphuric acid, leaving a residue of glistening scales of silica. Biotite is the second most important mica.

Lepidomelane is an iron-potash mica. It occurs in small six-sided tabular crystals, or in aggregations of minute scales. Color — black; luster — adamantine or somewhat vitreous. It is easily decomposed by hydrochloric acid, leaving a fine, scaly residue of silica.

Lepidolite, or lithia mica, resembles muscovite in crystalline form and many of its physical properties. Its color is white, yellowish or rose-red, the last being most common. It fuses before the blowpipe more readily than muscovite, and is decomposed by hydrochloric and sulphuric acids, but not so readily as the magnesian micas. Lepidolite is mostly met with in metalliferous veins, especially those containing tin, and is nearly always associated with other minerals that contain fluorine, such as fluorspar, topaz, tourmaline, and the emerald; it is also frequent in many kinds of granite.

Amphibole, often called **hornblende.** The prevailing

kind is an iron-bearing variety, in black cleavable grains or oblong black prisms cleaving longitudinally in two directions inclined to one another at 124° 30′. It occurs also in distinct prisms of this angle, and in all colors from black to green and white.

Actinolite is the name applied to the green variety, containing iron in addition to lime and magnesia. It occurs often in fibrous or columnar masses, sometimes with a radiated structure.

Tremolite is a lime-magnesia hornblende. The pure crystals are white, but the impure ones are yellowish or greenish-gray owing to the presence of protoxide of iron. There are several varieties of tremolite, as follows:

Asbestos * is in most cases tremolite containing a little water. It generally occurs in fine fibers which may be isolated or packed closely together with their principal axes parallel.

Mountain leather is a similar mineral, but the fibers are finer, closer, and intermixed.

Mountain cork is a spongy, elastic asbestos, with the fibers interlaced. **Mountain wood** is similar, but denser, far less elastic, and capable of taking a high polish.

Nephrite or oriental **jade** is a compact variety much used by the Chinese as a figure stone. The color is sometimes light-green as in the **white jade**; and olive-green, as in the **green jade**. It has an uneven, fine-grained fracture, and a greasy luster.

Tremolite is found in many places, but nearly always in the older dolomites and saccharoidal limestones.

---

* Most of the asbestos mined is a fibrous variety of serpentine, and is easily distinguishable because it contains about 14% of water.

Pyroxene, including augite. Like hornblende in most of its characteristics, its variety of colors, and its chemical composition. But the crystals, instead of being prisms of 124° 30′, are 87° 5′. Black and dark-green pyroxene in short crystals is called **augite.** It is an iron-bearing variety, and is common in igneous rocks.

The minerals of the amphibole group closely resemble pyroxene in chemical composition; they also crystallize in the same system. They differ, however, in the angular measurements of the oblique rhombic prism, which, as already shown, in hornblende is 124° 30′, and in augite 87° 5′ to 92° 55′.

They are all bi-silicates of protoxides and sesqui-oxides, the former being lime, magnesia, soda, potash, and the protoxides of iron and manganese; while the latter are represented by alumina and the sesqui-oxides of iron and manganese.

Crystals of amphibole differ from those of pyroxene, not merely in the angular measurements of their oblique rhombic prisms, but also in the angles at which their cleavage planes intersect. This circumstance is of considerable value to the mineralogist, since it is often difficult or impossible to measure the angles of the actual crystallographic faces, but it is generally possible to measure the angles of cleavage. The crystals of minerals belonging to the amphibole group usually exhibit a fine longitudinal striation.

Color affords no safe means of discriminating between pyroxene and amphibole, since the members of both groups exhibit greenish and brownish tints. The augites and hornblendes that occur in basalt are mostly brownish in color.

The hornblende in syenite is also usually brown, but that which occurs in phonolite is mostly of a greenish tint; while the augite in leucite lavas (these contain potash) is, as a rule, also green.

The minerals of the amphibole group commonly show a tendency to develop long, blade-like crystals. This tendency is in a marked degree shown by actinolite, one of the principal varieties of amphibole, the crystals arranging themselves in radio groups.

Both hornblende and augite occur together in the same rock. As a rule the former mineral is found in those rocks that contain a large amount of silica, the associated minerals usually being quartz and orthoclase; while augite is generally found in rocks of a basic character containing triclinic feldspars, and with little or no free silica.

Chlorite occurs sometimes in thin, foliated plates like mica, but inelastic, more often granular, or massive; sometimes in green crystals and scales. These kinds of chlorite are found in rocks, and form the mass of chlorite rock and chlorite slate.

The chlorites are silicates of alumina, iron, and magnesia with water, the average percentage of magnesia being 34, and that of water over 12.

Chlorite is a very soft mineral, and is essentially a product of the decomposition of other minerals.. When heated in a glass tube it gives off water. Before the blowpipe it exfoliates, whitens, and melts with difficulty into a grayish enamel. It is soluble in hydrochloric acid when powdered, and after long boiling.

Talc. A hydrated silicate of magnesia from which the water is only driven off at a high temperature. It usually occurs in broad, pale-green, or silvery-whitish plates or leaves, looking like mica; but the cleaved plates, though flexible, are much softer and not elastic. It is easily scratched by the nail, has a pearly luster, and is soapy and unctuous to the touch. Before the blowpipe it turns white and exfoliates. It is neither before or after ignition soluble in either hydrochloric or sulphuric acid, thus differing from chlorite.

Serpentine. This is also a hydrated silicate of magnesia. It is usually compact, massive, not granular at all, of a dark-green color, but varying from pale green to greenish-black. The most peculiar variety is a fibrous kind occurring in seams in massive serpentine, which is called **crysotile,** popularly called **asbestos.**

Minerals are composed of chemical elements, which are substances that cannot be further separated. A table of the chemical elements, their symbols, and equivalents will be found in the Appendix. When these elements unite and form a compound, they always do so in fixed proportion and in definite weight. Therefore, in any pure mineral, the composition of which is known, the amounts of the elements going to make up any given mass of it can be calculated readily.

For example, in galena (PbS) we have lead (Pb) = 207 and sulphur (S) = 32, total 239. Therefore, in 239 lb. of pure galena we will find 207 lb. of lead ($86\frac{1}{2}\%$), and so on in proportion.

Thus, any mineral that is pure enough to be weighed directly, or that can be concentrated pure and then weighed, can be estimated in this way, and the percentage content of the elements calculated.

The combination of two or more elements produces three classes of substances, namely, **acids, bases, and salts.**

Oxides of non-metallic elements are acids.

Oxides of metallic elements are bases.

Where an acid and a base unite, one exactly neutralizing the other, a substance is produced having neither acid nor basic tendency. It is known as a **salt.**

Most minerals are salts. There is only one common acid mineral, namely, quartz (SiO_2), or the oxide of the non-metallic element silicon.

There are many minerals that are basic, such as hematite (Fe_2O_3) and magnetite (Fe_3O_4), the oxides of iron, and cuprite (CuO), the oxide of copper.

Among the many mineral salts are: common salt or sodium chloride ($NaCl$); limestone, or calcite ($CaCO_3$), formed from the union of the oxide of calcium (metal) and carbonic acid gas; gypsum ($CaSO_4$ $2H_2O$ formed by the union of the oxide of calcium (metal) and sulphuric acid; apatite, a phosphate of lime [$Ca_3(P_2O_4)_2$] formed by the same base as above, uniting with phosphoric acid.

There are a great many minerals, the acid member of which is silica, with one or more metallic oxides forming the basic member. These are known as **silicates,** of which feldspar, mica, hornblende, pyroxene, talc, and . serpentine, are examples.

These facts are important to remember, because whole families of minerals and rocks are classified as acid or basic, according to the greater or lesser quantity of silica present in them.

Colors of minerals are either inherent, as in the sulphides, oxides, and acid compounds of most metals, and in those species of which they are essential constituents; or they are the effect of casual inter-mixture of these substances in species which, when pure, are naturally colorless. Of the latter sort are the colors of feldspar, calcspar, rock salt, marble, and jasper, in which the various tints of red and yellow are generally due to the oxide and hydrous oxide of iron. Other minerals derive a brilliant green color, some from carbonate of copper, others from the oxide of nickel or of chrome. Sometimes the intensity of the color is so far varied by a difference of texture or confused crystallization, that red, brown, and green substances appear, in a mass, to be black; but on being pulverized, their true color appears. It is therefore advisable, in describing a mineral, to state what its color is when reduced to powder.

The mechanical inter-mixture of coloring matter often renders a mineral more or less opaque; thus the red and yellow jasper are chalcedony — which, when pure, is highly translucent, or even semi-transparent — colored by minute particles of oxide of iron, which are themselves opaque. But colors which, though they may not be essential to a species, are the result of chemical combination, do not impair its transparency. This is illustrated in the violet tint of amethyst, which is derived from a minute portion of the oxide of manganese combined with the quartz; and the green of the emerald, which may in some cases be due to oxide of chromium.

Unequally distributed colors often produce parallel bands, either straight or curved, and clouded forms, as in agates. Sometimes the color takes the form of leaves and moss, or runs through the mass in veins, as in marble.

Double refraction gives rise to a phenomenon termed **polychroism.** Some minerals, placed between the eye and the light, transmit different colors in different directions. Tourmalines, viewed parallel to their axes, are generally opaque; perpendicular to them, they appear to be green, red, brown, etc. This difference is not observable in all double-refracting substances; but in some that have two axes of double-refraction, three different tints have been observed. Minerals crystallizing in the cubic system never transmit more than one color, if their composition and texture be homogeneous throughout.

In some minerals a peculiar light, termed, 'phosphorescence,' is produced either by friction or heat. On rubbing together two fragments or pebbles of quartz, a faint greenish light will be perceived; the same effect can be produced with certain marbles. Other substances, when placed on a heated shovel, emit a brilliant phosphorescence, which in some is green; in others pale violet. The best mode of conducting this experiment, if the specimen is powdered, or in small fragments, is to strew it over a shovel heated nearly to redness; but if it be an inch or two in length, it is better to heat it slowly and not beyond the necessary degree, by which means the operation may be frequently repeated without injuring the specimen.

Some metals are found native and in some degree of purity, as in the case of gold, silver, copper, mercury, and platinum. When so found they are easily recognizable. But frequently native metals appear under such colors, and even forms, that their appearance is deceptive. Gold, as an illustration, is frequently found in various shades of yellow, in accordance with the amount of silver

or copper it may contain, and yet to the practiced eye of
a true mineralogist it never loses the true gold hue.

Iron pyrite, which is composed of sulphur and iron,
and called ' pyrite,' mineralogically, sometimes has a color
somewhat similar to that of gold (called ' new chum gold '
in Australia) ; so also has a mineral called ' chalcopyrite,'
or copper pyrite, which contains copper, iron, and sulphur.
These, with others, vary in the yellow shade and degrees
of color, but are instantly detected by the practiced eye.
Of course the brittleness of these minerals is unlike the
softness of native gold, and this would instantly reveal the
fact that they were not gold; but we are now speaking
of the practiced eye alone, and therefore of the benefit of
cultivating a knowledge of minerals by sight.   The mode
in which a mineral breaks when struck smartly with a
hammer, or pressed with the point of a knife, is a characteristic of importance.   Many minerals can only be
broken in certain directions ; for instance, a crystal of calc-
spar can only be split parallel to the faces of rhombo-
hedron.   Many crystals break more readily in one direction than in others.   Whenever a mineral breaks with a
smooth, flat, even surface, it is said to exhibit

Cleavage, which always depends upon the crystalline
form.   But minerals often break in irregular directions,
having no connection whatever with the crystalline form,
and this kind of breaking is called

Fracture.   The nature of the surface produced by fracture is often of importance.   Thus quartz and many mineral species show a shell-like fracture-surface that is
called ' conchoidal,' or if less distinct ' sub-conchoidal.'
More commonly the fracture is simply said to be ' uneven,' when the surface is rough and irregular.   Occa-

sionally it is 'hackly,' like a piece of fractured iron. 'Earthy' and 'splintery' are other terms sometimes used and readily understood.

Streak. The color and appearance of a furrow on the surface of a mineral, when scratched or rubbed, is called the streak. This is best obtained by means of a hard-tempered knife or a file. The color of a mineral and its streak may correspond, or may show different colors, or the mineral may be colored while its streak is colorless. For instance: cinnabar has both a red color and a red streak; specular iron has a black color, but a red streak; sapphire has a blue color, but a white, colorless streak. The streak of most minerals is dull and pulverulent, but a few exhibit a shining streak like that formed by scratching lead or copper. This kind of streak is distinguished by the term 'metallic.' In judging the streak of a mineral, weathered samples should be rejected.

Hardness is another characteristic of great importance in identifying minerals; it is the quality of resisting abrasion and fracture. The diamond is the hardest substance known, as it will scratch all others; talc is one of the softest minerals. Other minerals possess intermediate degrees of hardness. To express the relative hardness of any mineral, it is necessary to compare it with some known standard, so 10 degrees of hardness have been chosen, as under:

1. **Talc** — easily scratched by the finger-nail.

2. **Gypsum** — does not easily yield to the finger-nail, nor will it scratch a copper coin.

3. **Calcite** — scratches a copper coin, but is also scratched by a copper coin.

4. **Fluorite** — is not scratched by a copper coin, and does not scratch glass.

5. **Apatite** — scratches glass with difficulty; is readily scratched by a knife.

6. **Feldspar** — scratches glass with ease; is difficult to scratch with a knife, but is scratched by a well-tempered steel.

7. **Quartz** — cannot be scratched by a knife, and readily scratches glass.

8. **Topaz**
9. **Corundum** } harder than flint or quartz

10. **Diamond** — scratches any substance.

The metal tantalum is considered to be almost as hard as the diamond.

In describing minerals, their hardness is always expressed by numbers. Thus, if on drawing a knife across a mineral it is impressed as easily as calcite, its hardness is said to be H3. If a mineral scratches quartz, but is itself scratched by topaz its hardness is between 7 and 8.

In testing a mineral for hardness, a sound portion should be chosen, and the scratch made on a smooth, clear surface, and with a sharp edge or angle of the mineral used for scratching. On scratching one mineral with another there may be a streak of dust, which may come from the waste of either, so it cannot be determined which is the softer rock until the dust is wiped off, when it will be easily seen that no scratch has been produced on the harder mineral, and that the edge of the other has been blunted. This is what would happen if an attempt were made to scratch topaz with quartz, or corundum with topaz.

From the test of hardness, clear distinctions may be drawn between minerals that resemble one another. Iron pyrite and copper pyrite, for instance, are similar in appearance, but the latter can easily be scratched with a knife, while the former is nearly as hard as quartz and the knife makes no impression upon it.

Flexibility and elasticity. Some minerals can be readily bent without breaking, for instance, talc, mica, chlorite, molybdenite, native silver, etc. Minerals, which after being bent can resume their former shape, are called elastic; mica and elaterite are examples. A remarkable instance of flexibility, even combined with elasticity, among the rocks, is that of a micaceous sandstone called itacolumite, which, in Brazil, is the matrix of the diamond.

Smell. Only a few minerals, like bitumen, have a strong smell which is readily recognized, but specimens generally require to be struck with a hammer, rubbed, or breathed upon before any smell can be observed. Some black limestones have a bituminous odor, while some have a sulphurous, and others a fetid, smell. Hydraulic limestone has a smell of clay, which can be detected when the mineral is breathed upon. Some minerals containing much arsenic, mispickel, for instance, smell of garlic when struck with a hammer. Specimens of rocks kept in show-cases for any time emit a general odor of chemicals when opened up. This is probably due to some decomposition going on.

Taste. Only soluble minerals have any taste, and this can only be described by comparison with well-known substances, for instance, acid, vitriol; pungent, sal-ammoniac; salt, rock salt; cooling, nitrate; astringent, alum;

metallic astringent, sulphate of copper; bitter, sulphate of magnesia; sweet, borax.

Malleability. Substances that can be hammered without breaking come under this heading. It is on this quality that the value of copper, silver, gold, iron, etc., in the arts depends.

A few minerals are malleable, at the same time sectile, that is, they can be cut with a knife; silver glance, horn silver, and ozokerite are examples.

Mineral caoutchouc (elaterite) is sectile, but like india-rubber, can only be shaped when hot. The elasticity of elaterite is so characteristic that the mineral will be readily recognized.

Ductility, or the capability of being drawn into wire, is a property confined to a few metals. It is possessed in the highest degree by gold, which can be drawn into the finest wire, or hammered into leaves of such fineness that 30,000 of them are not thicker than an eighth of an inch. Aluminum, copper, iron, and tantalum are also drawn into wire, especially the last named, for electric lamps.

Luster. This term is employed to describe the brilliancy or gloss of any substance. In describing luster, well-known substances are taken as types, and such terms as 'adamantine luster' — diamond-like — and 'vitreous luster' — glassy — are used. The luster of a mineral is quite independent of its color. When minerals do not possess any luster at all they are described as 'dull.' The kinds of luster distinguished are as follows:

Metallic: The luster of a metallic surface, as of steel, lead, tin, copper, and gold.

Vitreous or glassy luster: That of a piece of broken glass. This is the luster of most quartz, and of a large part of non-metallic minerals.

Adamantine: This is the luster of the diamond. It is the brilliant, almost oily, luster shown by some very hard materials, such as diamond, corundum, etc. When sub-metallic it is termed 'metallic adamantine,' as seen in some varieties of lead ore like cerussite.

Resinous or waxy: The luster of a piece of rosin, as that of zinc blende, some varieties of opal, etc. Near this, but quite distinct, is the 'greasy luster,' shown by some specimens of milky quartz.

Pearly or the luster of mother-of-pearl: This is common where a mineral has perfect cleavage. Talc and native magnesia are examples.

Silky, like silk: This is the result of fibrous structure, as the variety of calcite (or of gypsum) called satin spar, also of most asbestos.

Fusibility. Some minerals can be fused easily; others only with difficulty; while some resist the highest heat. There are such wide differences between the degrees of fusibility of minerals that this characteristic helps greatly in identifying them. Fusibility is most readily tested by holding a small splinter of the mineral with a forceps in a candle flame, urged by the blowpipe; or the mineral may be laid upon a piece of charcoal and the flame directed upon it by the blowpipe. Some minerals fly to pieces when heated; others swell up or give off peculiar and characteristic odors. For further information regarding fusibility, see Chapter II, 'The Blowpipe and its Uses.'

Specific Gravity. By specific gravity is meant the

comparative weight of equal bulks. Water is taken as the standard of comparison; the specific gravity of a mineral is a number showing how many times it is, bulk for bulk, heavier than water.

The specific gravity of water is called 1, and of gold 19, implying that if equal bulks (or volumes) of gold and water were taken, the gold would weigh 19 times as heavy as the water. The specific gravity of a mineral can be found by weighing it first in air in the usual manner, and then observing how much of its weight it loses when suspended from the arm or pan of a balance, and allowed to hang freely in water. If a piece of quartz weighing 26 grains is attached by a horse hair or fine silk thread to the scales — and weighed while hanging in water — it will be found to weigh only 16 grains; it thus loses 10 grains, or $\frac{10}{26}$ of its entire weight, and its specific gravity is 2.6, a figure obtained by dividing the weight of the mineral in air by its loss of weight in water. Similarly gold would lose $\frac{1}{19}$ of its weight.

Minerals differ widely in the proportion of weight that they lose in water, but the same mineral invariably loses the same proportion, for instance: quartz loses $\frac{10}{26}$ of its weight; topaz, $\frac{10}{35}$; sapphire, $\frac{10}{40}$; zircon, $\frac{10}{45}$; and tin ore, $\frac{10}{70}$.

These proportions depend upon the specific gravity of these minerals. The specific gravity of water is called 1, of quartz, 2.6; of topaz, 3.5, of sapphire, 4; of gold, 19.

In determining how much weight a mineral loses in water, a delicate balance is required when the weight in air is under 10 grains; but for portions weighing heavier than this, a common balance, sensitive to a grain, may be used for practical purposes. The mineral must be sound, clear of foreign matter, free from pores or cracks, and its surface should be rubbed with water before immersing it, so as to prevent bubbles of air adhering, which would affect the result.

Some rules for finding weights by specific gravity are given in the Appendix.

While the specific gravity of a mineral may be ascertained with great accuracy in the laboratory where delicate balances are available, it is not always possible to do so in the field, and the most that can be undertaken is to class minerals roughly within certain broad limits. Prospectors soon acquire some proficiency in testing the weight of minerals by handling them. A lump of pyrite, for instance, can readily be distinguished from gold by its weight, since a mass of gold of the same size would weigh at least three times as much.

A rough idea of the specific gravity of minerals can be arrived at by washing in a tin dish. In the hands of an experienced man this method will give results accurate enough for the determination of the commonest minerals. Sorting in a dish is effected by picking out the larger stones by hand, but in testing the specific gravity of minerals, they should be divided, in the first instance, into regular sizes by screening. For this purpose two sieves will be sufficient, one with eight holes, the other with sixteen holes to the linear inch; then all that will pass through the coarser sieve, but not through the finer, will be of sufficiently uniform size for the tests required.

The lighter portion will first be separated by washing; these will consist of shale, ferruginous quartz, brown oxide of iron, pebbles of tourmaline, etc., mostly of a lower specific gravity than 3.5, and the heavier minerals that remain in the dish, will be zinc blende, magnetite, pyrite, hematite, mispickel, tinstone, wolfram, gold, platinum, etc.

By careful manipulation of the dish, generally practiced by miners when showing the gold, these heavy minerals can be easily enough separated into three groups, namely: gold, platinum, copper, bismuth, silver, mercury, etc.; tinstone, wolfram, galena, cinnabar, etc.; and zinc blende, magnetite, hematite, mispickel, etc.

Some of these minerals — mispickel for instance — can be readily recognized. Where this is the case, those that lie 'upstream' and those below can be sub-divided as being of greater or less respectively than 6.3, which is the specific gravity of mispickel. Where the minerals in the dish cannot be readily identified, a few fragments of metallic antimony, or zinc, or tinstone painted white — all of which have a specific gravity of about 7 — should be introduced into the dish to serve as a gauge.

What has previously been said of color may also be said of 'weight' and 'form.' A piece of pyrite would not be mistaken for gold on account of its weight. Three crystalline pieces — one of barite and the other two of lime carbonate and of quartz — may appear equally transparent to the unskilled eye; but the form of the first is tabular, and that of the latter two is in six-sided crystals, the lime carbonate crystals terminating in three sides, and the quartz always (like the sides) in six.

Besides a knowledge of the forms of minerals, it is also necessary to learn the characteristics of some of the rocks that are generally associated with those minerals. This is essential for intelligent field work. For instance, the iron ores — brown hematite and black band ore — would not be expected in granite country, yet magnetic ore and red hematite might be found there. It is, therefore, important that the prospector know many of the rocks to help in guiding or in checking him in his explorations. A general knowledge of the manner in which the

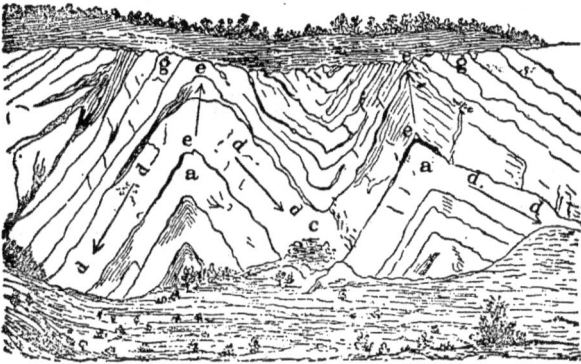

FIG. I.— SECTION SHOWING CONTORTED STRATA DUE TO
LATERAL PRESSURE:

(aa), anticlinal axis; (c), the synclinal axis. The direction of the arrows (ee, ee), is that of the strike. That of the arrows (dd), is that of the dip of the strata, always measured from the horizon; (gg), are the outcrops.

rocks were ' laid down,' their order, or succession, in the earth, is important, and the distinction between sedimentary and that which has been, and is usually called ' igneous rock,' but more properly 'azoic rock,' that is, rock which does not exhibit any remains of fossil or organic life. Often the only signs by which one can, with any degree of certainty, determine what is the name of the sedimentary rock, is by finding the remains of former life, that is, the kind of fossil it contains. Dana said that it is settled that the kind of rock in itself considered is not a safe criterion of geologic age.

If all the rocks in the world had been laid down in regularly horizontal sequence and had always remained in their own separate ' horizons,' as every rock of the same age is called, not only should they be found all parallel, one over the other, but we might readily determine to some extent what were the exact order and distance of any one horizon, or geologic age. Although there is a general order, the same in all parts of the world, there have been upheavals and sinkings, dislocations and erosions, so it is necessary that the prospector should become acquainted with the various probable changes in the order and forms of the rocks that carry the minerals he is seeking. Some of the movements of the earth's crust are shown in Fig. 1.

PRACTICAL GEOLOGY.

It has been mentioned that useful minerals are not always confined to one horizon, but there are certain ranges of rock that indicate their vicinity. There are also limits that are never passed by some useful minerals, and experience has shown that some horizons are always barren of ores, and it is therefore useless to expect to find them in commercial quantities in certain rocks or beyond them in certain directions.

Gold is often found where it will not pay to open and work the strata; so are lead and copper. It is well to learn the relations of such barren regions, or horizons, as the strata are called. In the table following chief place has been given to those horizons that have been found in the United States to abound in the useful minerals. ·Small specimens of the principal rocks mentioned should be procured and examined under a good lens, so as to become thoroughly acquainted with them.

All rocks may be classed as

(1) igneous, (2) metamorphic, and (3) aqueous.

A rock may be defined as a mineral aggregate possessing a more or less persistent geologic character. However, speaking geologically, not only the hard consolidated massive and stony substances are called ' rocks,' but any natural deposits of stony material such as sand, earth, or clay, when in natural beds, are geologic rocks. Few of the rocks of the earth, at any rate so far as examined, are in their original and primal condition. Even the granites and volcanic rocks are composed of other and more ancient material disintegrated, ground up, or worn down, settled, buried, and compressed by ages of enormous pressure, or consolidated by cementation. Some have been ' laid down ' under water, having been disintegrated into dust, carried by winds over the oceans, and settled down into the form of the present rocks, which afterward have been lifted up into mountains and plains above the seas. The transporting power of rivers or currents in ancient oceans, and unequal upheaval of some regions where subterranean forces were greater than at distant places, have caused large differences in the nature of the deposit, even in limited areas. These special and limited forces will account for the fact that although, taking the geologic horizons throughout the world, there is a general sameness, and important members of the order of succession are omitted in some regions, and exceptions to general rules occur.

In the table following therefore are given those universally accepted relations of certain rocks, one to another, in the great geologic arrangement of the earth, omitting some of the subsidiary, limited, and unimportant horizons.

1. **Igneous Rocks** are such as owe their origin to the action of fire, having been subjected to sufficient heat to melt the ingredients. They form the smaller, but still a large, part of the crust of the earth. They are not sedimentary, but are due to upheaval. They are not stratified and not fossiliferous. Some geologists divide them into ' plutonic ' and ' volcaniç ' rocks, the former being crystalline, older, deeper in origin; the latter noncrystalline, and comparatively recent and surficial. In the case of the plutonic rocks, the rate of cooling has been slow, the consolidation gradual, and has taken place under great pressure. The rate of cooling of the volcanic rocks has on the other hand been fast, and the consolidation rapid, as with lava, etc.

Trachyte is grayish rock of rough fracture ; the same specific gravity as quartz, but mainly constituted of grains of glassy feldspar: It is essentially a uni-silicate of alumina, with 10 to 15% potash, a little soda and lime ; it differs from quartz in that it fuses before the blowpipe, while quartz remains unfused, except when soda is used.

Basalt is blackish or dark brown. Traps, greenstone, dolerite, and amygdolite are only modifications, being all uni-silicates with smaller amounts
of potash than in trachyte, a little more soda and
lime, and some traces of iron and magnesia, varying in color and form.

Obsidian is a volcanic glass, something like bottle
glass, of a dark shade, and translucent.

All of these are compact in texture except where holes
have been caused by steam or gas. They are frequently
found penetrating several strata, having been forced up
in columns almost vertically, and sometimes spreading
out horizontally for many miles between the strata or
on the surface, and are called volcanic dikes, or intrusive
rocks or lava.

The granite rocks are of igneous origin. When they
reach the surface they are termed eruptives.

2. **Metamorphic Rocks.** The term 'metamorphic' as
applied to these rocks, implies that they are the
product of the metamorphosis of rocks originally sedimentary. They are of igneous origin, subsequently to the time when they were of aqueous
origin, and have undergone a change through pressure and heat, perhaps in connection with steam
and water. All the rocks òf this class are to be
identified from the igneous by their foliated texture; yet more by their alternate bedding in parallel layers or strata, and the distinct traces that
they often show of internal stratification. All
the metamorphic rocks are silicates and acid silicates. They contain from 42 to 75% of silica.

All carry aluminium, magnesium, iron, calcium,
in the above quantitative order, and all but talc-
schist contain small quantities of potassium and
sodium. Of this class of rocks are the following:

Gneiss, having a composition of small pieces of
feldspar, mica, and quartz, like some granites, but
laminated or foliated in form, and not equally
solid, homogeneous, and continuous throughout its
structure as granite is

Mica-Schist. This term is given to those laminated rocks composed of mica and quartz in small
particles, easily broken up, but more easily broken
into tabular or leaf-like pieces, because the mica
has been deposited in planes allowing of cleavage.

The principal rock of Manhattan Island, upon which
the city of New York stands, is a schist-gneiss. At the
south end it is often covered by sand, gravel, and hard
pan; but towards the north and on the mainland this rock
outcrops prominently, frequently containing quartz veins.
Nearby, in Jersey, is much sandstone, used largely in
building.

3. **Aqueous Rocks** are simple water rocks — that is,
rocks composed of sediments from the dust or
ground-up remains of other rocks. The presence
of such sediments is due to the transportation
power of rivers, floods, or currents, also of winds
and other agencies, carrying the dust to the oceans
where it was arrested and became a sediment.

These rocks are stratified, as a rule, or made up of
successively deposited strata; and are mostly fossiliferous.

STRATIFIED ROCKS.

GENERAL DIVISIONS.	SUB-DIVISIONS.	CHARACTERISTICS.
RECENT, PLEISTOCENE, OR QUATERNARY.	All its shells and bones are of existing species.	Tertiary rocks, yield brick and other clays, gypsum, sand, phosphate of lime deposits such as are in Florida, South Carolina, and elsewhere; gold in the drift and alluvial, also platinum (iridium, *see text*), and tin.
PLIOCENE.	About 50% of existing species of shells.	Coal fields (brown or lignite) of this period, occur in India, Indian Archipelago, Japan, New Zealand, Vancouver Island, and in Europe; also in California, Washington, Oregon, Colorado, etc.
MIOCENE.	Contains 80% of extinct species.	The true coal (anthracite and bituminous) belongs to the Carboniferous only.
EOCENE.	Contains fresh water and marine strata, animals all extinct.	A hard lignite exists at Gay Head, Martha's Vineyard, in this formation.
CRETACEOUS.	Upper. / Middle. / Lower.	Upper chalk, with flints, but the lower chalk without flints. } The whole formation contains sea-shells, sponges, sea-urchins, etc. Contains greensand in England and in New Jersey, used as a marl and fertilizer. There is a supposed cretaceous lignite in Alaska, Colorado, California, Utah, etc.
JURASSIC.	Whealden. / Portland stone. Oxford group. Stonesfield slate. / Lias. Limestone in horizontal strata.	Consists of sand, clay, or marl, the sand used in glass-making. Some British coal is found in the Oolite; Kimmeridge clay is found in upper Oolite; the fine Bavarian lithographic stone in the middle Oolite.
TRIASSIC.	Keuper. Muschelkalk. Bunter-sandstone.	Conspicuous for the number of ammonites and nautilus shells. Furnishes building and paving stone. Called by the Germans, 'trias.' Connecticut river sandstone with footprints. The new red sandstone of England. Red clays, marls, shales, and sandstones. In Europe great salt beds.

CENOZOIC. OR TERTIARY.

MESOZOIC. OR SECONDARY.

PRIMARY OR PALEOZOIC.			
PERMIAN.	Dark red sandstone. Magnesian limestone. Conglomerates, breccias, marls in all three.	Mostly sandstones and marlytes, some impure magnesian limestone and gypsum. Thin seams of coal unworkable. With exception of brown hematite iron ore and the metals mentioned above, all the other metals are found in the formations below.
CARBONIFEROUS.	Seams of anthracite and bituminous coals of varying thicknesses. Millstone grit. Sub-carboniferous.	The black band iron ore. Limestone from the same mines with the coal in Great Britain, but not so frequently in America. Anthracite, cannel, and bituminous coal in seams in limestone, sandstone, and shales, forming the coal measures.
DEVONIAN.	Catskill period. Chemung period. Hamilton period. Corniferous period.	Affords petroleum in Pennsylvania, Ohio and elsewhere, and salines in Michigan. It is the mountain limestone of England. Largely of corals. Includes the old red sandstone of England. Hamilton black shales produce oil, the Hamilton beds afford excellent flagging stone. Corniferous called also upper Helderberg group.
SILURIAN. Upper	Oriskany sandstone. Lower Helderberg period. Salina period. Niagara period.	Salina period supplies the salt waters of Salina and Syracuse, N. Y.
SILURIAN. Lower	Trenton period. Canadian period. Potsdam sandstone.	The lead mines of Iowa and Wisconsin are in the magnesian limestone of the Canadian period.
	Cambrian. Laurentian. ARCHÆAN.	

The grains of sand in sandstone are rounded, having no sharp edges as in granite.

Where the sedimentary material was exceedingly dust-like, it sometimes is laid down as fine mud and frequently in laminæ, as in shale.

Granite is a term descriptive of rocks generally composed of quartz, feldspar, and mica, in grains (hence the name) of a crystalline form. But the granites are not all alike in the amount of either of these minerals, nor are they alike in color. Some granites contain no mica, as in 'graphic' granite, but only quartz and feldspar, the former in the feldspar resembling written characters. Others contain hornblende as well as mica, or in place of mica, the hornblende being in dark or black crystalline specks, pieces, or crystals, and consisting essentially of silica, magnesia, lime, and iron. This granite is called 'syenite' granite. Where the feldspar is in distinct crystals in compact base, and sometimes lighter than the base, which is frequently reddish, purple, or dark green, it is a 'porphyritic' granite. The granites are sometimes whitish, grayish, or flesh-red. They are considered as metamorphic and not igneous (Dana), although some mineralogists still consider them to be igneous. They always present a crystalline grain in varying degrees of fineness and prominence.

Some specimens of this rock contain two kinds of mica, one black — biotite — the other white of silvery appearance, muscovite. The biotite presents in spots the appearance of hornblende, and only the pen-knife point shows the scaly lamination of mica under the lens. It also contains crystalline forms of potash feldspar (orthoclase), distinguishable from the quartz by their sides only, by the lamellar fracture of its edges, and its peculiar vitreous glimmer, as the hardness appears practically the same, although feldspar (6.6 and quartz 7) is slightly softer. It would be well for the prospector to gather many forms of granite and examine them under the lens until he becomes thoroughly acquainted with the variations.

Since their original formation, rocks have been subjected to numerous changes. Some have been raised from the sea without being lifted to any extent from their original horizontal position; others have been folded into most fantastic shapes; while others again have been completely inverted. In other cases, movements that have taken place since the rocks became solidified have caused fractures, and by the rocks on one side of the crack sliding on those of the other 'faults' have been produced.

Where the rocks have been folded in the shape of an arch, they are said to form an 'anticline,' and where they occupy a basin, they are spoken of as forming a 'syncline.' One is the reverse of the other.

In examining the surface of a country in which the rocks are of sedimentary origin, it will be found, as a rule, that the beds are inclined at varying angles to the horizon, and in making a geologic survey of any special district it is necessary to note the 'strike' of the rocks at every available point. When any well-marked bed occurs, its line of outcrop should be carefully followed and mapped; the boundaries of any eruptive rocks should also be clearly delineated on the plan.

The strike of a rock is the direction of a horizontal line in any of the beds, or, in other words, the direction

in which a level drift would be put in on the floor of the
bed. The ' dip ' is a line at right angles to the strike on
the plane of the beds, and the angle is to be measured in
relation to the horizon.

When any particular bed is followed on the surface,
it is often found that it does not continue with the same
strike for any great distance; in fact, it gradually veers
round, the direction of the dip changing at the same time.
The boundaries of rocks are sometimes rather obscure
in consequence of the variable movements that have taken
place, but tracing them on the surface is made most difficult by faults and dikes. The displacement due to faults
may be only an inch or so, or may be several hundred
feet (in one mine in California a lode was faulted 1100
feet), while in exceptional cases it may be as much as two
or even three miles. A study of faults is of great importance, more especially on account of their close association with mineral lodes. They are common in most
mining districts, and often result in the loss of a lode or
it loses its mineralization.

The first indications of a deposit possessing economic
value are, as a rule, to be met with among the materials
forming the beds of streams, and wherever water-courses
have seamed and furrowed the rocks. Metalliferous deposits should generally be sought in hilly districts, though
alluvial (gravel) accumulations may be found in comparatively flat country. A close study of natural phe-
nomena will often help the discovery of minerals; for
instance, the form and color of the surface, stained
patches, springs of water whether sweet or mineralized,
scum floating on water (petroleum, etc.), accumulations
of earth brought to the surface by burrowing animals,

changes in vegetation, and behavior of the magnetic needle. These, however, only serve to indicate existence, without reference to quantity or quality.

The valuable minerals and metal-bearing deposits of the earth occur as

Lodes. When a fissure in the rocks of the earth is filled with mineralized matter it is called a ' lode' or ' vein.' In Australia a vein is often called a ' reef ' and in California a ' ledge.' Both are incorrect terms; a rock in the ocean is a reef, and a deposit of oil-shale is a ledge. The great gold deposits of the Witwatersrand in the Transvaal are called reefs. The course of a lode in a horizontal direction is called its strike or pitch, while its descent is spoken of as its dip. Lodes are often marked off from the rocks enclosing them by straight and sharp divisions on either side of the lode as if cut with a knife. These are called the ' walls.' When the lode inclines in its dip to either one side or the other, which is nearly always the case, the upper division is called the ' hanging wall,' and the lower the ' foot-wall.' The inclination of the lode in its dip is its ' underlie.' The barren rock through which the lode passes is known among mining men as ' country.' Veins may be all widths, from a streak half an inch to 100 feet or more. They often enclose large blocks of barren country rock or waste, and are spoken of as ' formations ' or ' horses.'

Lodes nearly always carry ' casing ' or ' gouge,' which is country rock and vein-rock ground fine. Casing is mostly found on the foot-wall, and is often rich.

That part of a vein which contains pay-ore is called an ' orebody ' or ' ore-shoot.' It usually has a vertical or diagonal dip on the plane of the dip of the vein. A

' lens ' is an orebody of lenticular form. The waste rock or mineral in which ore and metallic particles are held is called ' gangue.'

When a mineralized outcrop has been found and the strike, in case of a lode, has been determined, it is advisable to test it along the surface at various points to prove its continuity or persistence and richness. It must not be assumed because a lode is rich where discovered that it will be equally so at all points where it is opened; and conversely, because a vein is poor where first discovered, there is no reason to suppose that further prospecting along its course may not disclose parts in which there is pay-ore.

When surface prospecting has given as much information as possible, some sinking and driving should be undertaken to prove the persistence and value of the deposit at depth. To acquire the greatest amount of information at a minimum cost, the point for sinking a shaft should be selected on the surface where the vein is best and, having determined the extent along the strike — as nearly as possible — which carries payable mineral, the shaft should be placed about the center and sunk on the underlay to a depth of 100 feet, or less, if water is reached sooner; and from the bottom, levels should be driven along the course of the lode as long as the ore is of sufficient value.

It will be seen that ground may be opened cheaply in this manner, in which a certain quantity of ore can be measured and sampled, and an accurate idea of its value obtained. In measuring quartz, it is usual to estimate 13 cubic feet to the ton, in the solid, so that a vein 3 feet wide proved to a depth of 100 feet and for 100 feet along its ·strike would contain $\dfrac{100 \times 100 \times 3}{13} = 2307$ tons, provided that geologic conditions did not interfere with the deposit.

The ore should then be sampled in order to arrive at a fair estimate of its value.

Sampling ore is one of the most important operations in field work, in opening a prospect, and in the subsequent stages of developing a mine. In the past, most prospectors and others have been woefully ignorant of even the fundamentals of sampling. They forget that the future of their property depends upon fair samples being taken. How many times are pieces of rock and ore broken off haphazard, or from the best looking part of an outcrop or vein, sent to be determined mineralogically or for value, with eventual disappointment, in that when samples are well taken later on, the true condition is revealed. One or two pieces of rock is not representative of a whole deposit; an average is what is required. The same thing happens when lessees or tributers send gold ore to a custom mill for treatment. Probably they have not had any assays made, and rely on pan tests to guide them, which gives them a certain grade, but when the mill does not recover what they expect they allege poor treatment and that their gold is lost. The reviser of this book has seen many instances of this sort of thing, and no doubt the various Mining Bureau, Schools of Mines, custom works, and public assayers are often annoyed by unsystematic sampling. In the field, picks, hammers, moils, and gads should always be part of the equipment, and are useful in sampling; also an iron pestle and mortar. A piece of inch hexagon drill-steel, about 3 feet long and spread at one end to 2 inches diameter, makes a good pestle. A couple of panning-off dishes are also needed. Cleanliness is an essential in sampling, otherwise results are likely to be salted. In taking samples from an outcrop or face of an ore-shoot, a channel or groove 3 to 6 inches wide and from ½ to 1 inch deep should be cut across it. A fair quantity would be up to 5 pounds per foot. This should be caught in the hand, in a box, or on a piece of cloth. Any rock that flies to the ground should be discarded, as by picking it up the whole sample may be salted. If the face of a deposit shows rich streaks, it should be sampled in sections. Samples should be taken at distances no greater than 5 feet apart, unless a long and irregular outcrop is being examined, when the intervals may be 10 or 20 feet. While samples are being taken, the character of the rock should be noted at that point. To save freight or postage, large samples may be cut down in the field. This requires care. The pieces of ore are crushed in the mortar to say half-inch size, mixed on a clean piece of canvas or other material, spread out flat, cut into four quarters, the opposite ones being kept, these further reduced to say eighth-inch size and quartered, until a fair sample weighing a pound is secured. If the sample is to be panned, this remaining pound should be crushed to pass a 20-mesh screen. Panning and estimation of value of gold ore by such method requires much experience. Some men have acquired such proficiency in panning free-milling gold ore that they check assays very closely. The average value of a number of samples is not obtained by dividing the number into the total value. This is the arithmetical mean and is entirely wrong. The correct method is to multiply the width of each sample by its value, giving inch or foot-value or whatever are the metal contents, then adding the total foot-values and dividing by the total of all the widths, thus giving a true average or what is known as a geometrical mean. These remarks apply to any kind of ore. A good pocket lens should be carried, so that rocks being sampled or ore being panned may be examined.

In following minerals other than gold, it must be remembered that many of them have a tendency to decompose when exposed to the action of the weather and, consequently, that the nature of the ore at the outcrop may be very different to what will be found at depth. Copper ores, for instance, are liable to decompose, forming soluble sulphates, which are carried away in solution by running water. As most copper ores are associated with more or less iron, the outcrops of copper lodes are frequently represented by a porous ironstone which is called ' gossan,' no sign of copper being found until depth has been reached. Generally speaking, an outcrop of porous gossan may be looked upon as a good indication for mineral in depth, whereas a dense ironstone seldom leads to rich deposits of other mineral below.

The most common of bedded deposits are those of coal. Many kinds of iron ore are found in beds, also some copper ores in shale, silver and lead ore in sandstone, etc. Beds and layers are also known as ' strata,' ' measures,' ' sills,' ' mines,' ' bassets,' ' delfs,' ' girdles,' and ' seams.'

Irregular deposits consist of pockets, contacts, and stockwerks, where mineral is diffused through rocks, or in small cracks. Many irregular deposits are of great value, and some of the rarer minerals, such as sulphide of bismuth are found in them. Such deposits are not only irregular in their mode of occurrence, but vary considerably in size and shape, so that surface indications are no criterion as to their extent. It is more important in opening these orebodies to follow them better than a regular vein.

Surficial deposits. This term means beds of alluvium that more or less cover the surface of every country. These beds have been chiefly formed by various mechanical agencies, which, after having broken down the higher rocks, carry the material resulting therefrom down to lower levels. By this process most mineral deposits are so comminuted that by their exposure to the atmosphere they are decomposed and destroyed. However, substances like cassiterite (tin oxide), platinum, gold, etc., not being subject to decomposition, have, in consequence, been more or less preserved and buried among these surficial deposits. In studying these, note should be taken of their general situation, area, thickness, and richness. Several beds may be ranged one above the other, in which case their relative values have to be determined. In tracing any particular deposit, while ascending a valley or cañon, if the particles of ore increase in size and number, the prospector may expect that he is approaching their common origin. Another indication that he is near this point will be that he will find the mineral less worn.

Generally speaking, all metals are only found in the oldest rocks, which form the backbone, so to speak, of the main ranges of metal-producing countries. In selecting a region for prospecting, it will be best to seek one where the rocks are neither too hard nor too soft, nor should they be of too uniform a character. The country most deeply indented with gullies, cañons, and gulches running parallel to one another offers good chances. The region near the sources of the main rivers is frequently rich, and always the most easily prospected. Alluvial gold begins at or near the locality where a number of auriferous lodes exist in such regions,

When a river forks at its head into two or more branches, it is strange that the source of the gold will nearly always be found in the right-hand branch, which is the stream on the right when looking toward the mouth of the main river. This right-hand theory is an old mining superstition for which science has offered no explanation, but its almost unfailing applicability is fully established by practical experience.

The color of rocks also serves as a guide to prospecting. Rocks of a pinkish-reddish color alternating with rocks of a deep bluish tint streaked with drab are generally favorable for metallic deposits. Another good indication is when the faces of precipices are covered with a black ooze caused by manganese, the presence of which always indicates mineralization. These are simply general indications.

Although color is always a good guide to the location of metallic deposits, it is of special service to the prospector in unexplored districts. Thus copper is indicated by greenish, bluish, or reddish stains upon rocks in the neighborhood of the lode; tin and manganese by dull, black tints; manganese shows itself also in pinkish streaks. Gold, being always accompanied by iron, manifests its presence in red, yellow, or brown shades; lead and silver reveal grey or bluish-grey tinges; blende dyes the rocks yellowish-brown; and iron disports itself in all the hues of red, yellow-brown, and even dun-black.

The wash of rivers and creeks, and even more so that deposited upon terraces (if any) flanking the streams, must claim close attention. By wash is meant the diluvial drift in which gold or tin — the only metals mined in diluvial deposits — is found. Stones streaked with pinkish lines, indicating manganese, are always found in wash carrying gold. Green stones, which are universally found in the wash, are always a good indication of gold if they are of a bright sea-green or even pea-green, but they must be smooth, hard, well-polished, and heavy. In many districts such stones are considered the 'pilot stones' to gold. Quartz cobbles must be always present in considerable numbers (the gravel on the Sierras of California is 95% white quartz, a trifle stained brown) in all gold-bearing wash, and if showing signs of decomposition, are all the better as a favorable indication.

Much of the world's gold has been obtained from surficial deposits called 'placers.' Placer gold is always found adjacent to and lying below areas traversed by auriferous veins, and nowhere else. These veins were shattered and eroded in Tertiary times, which tended to break down and comminute the quartz and to liberate the gold therein.

Nuggets and coarse gold are found nearest the outcrops of the veins that have supplied them, while particles gradually become finer and finer as the line of drainage along streams is followed from this point.

Pebbles and fragments of gold-bearing quartz that have been derived from neighboring veins are commonly found in placer deposits, and most of the nuggets have more or less quartz, like that of the veins, still adhering to them. The gold is found in scales, grains, pebble-like nodules, and round, battered masses or nuggets.

Such alluvial deposits demand careful examination. They are of great importance because alluvial gold and tin are in many cases found under conditions that require no capital to work them and, consequently, immediate returns can be obtained when discovery has been made. River beds and creeks should be examined carefully, a pick and a shovel, a tin dish and a large knife being all the equipment necessary. In the first place the gravel should be washed carefully to see whether there is any gold. Then certain beaches along the stream should be selected and shallow pits sunk through them until bedrock is reached. All the material raised should be panned, bearing in mind that most of the gold is generally found on the bedrock.

A further test should be made by carefully following up the stream, especially when it is low, and cleaning out with a knife all crevices in the rocks in which gravel and sand have accumulated. This should all be panned. In some cases large quantities of gold have in a short time been saved by prospecting in this manner.

However, the gold that is found in rivers and streams does not necessarily point to the close proximity of the veins from which it was derived; still less does the occurrence of alluvial gold in buried river beds indicate the proximity of lodes. A careful prospector will often notice that a stream which he is investigating appears at certain points to have altered its course, having, in fact, found it easier to cut a channel in a different direction

to that which it originally followed. It did this leaving
its former channel, with the gravel and sand it had deposited, high and dry.   In cases such as this, it is generally worth-while to sink small shafts through the gravel
until bedrock is reached, and, if the first is not successful, others should be sunk towards the upper part of the
channel, as defined by the inclination of the bedrock
where it is encountered.   There are, of course, comparatively few prizes and many blanks in prospecting such as
this, but the value of the deposits found at times offers
inducements to continue trying, even when but small success has attended earlier efforts.

The domain of the prospector lies in hilly ground.
Flat plains have little attraction for him except under
special conditions, because, though valuable minerals may
be present, they are certain to be covered by a thick
blanket of soil.

In character, placers manifest almost as great variety
as vein deposits. The following illustrations show in section some forms of these alluvial deposits:

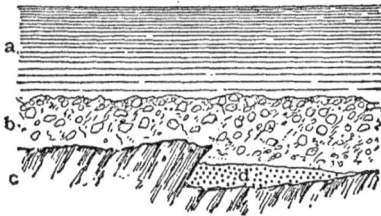

Fig. 2.

The stream (Fig. 2) flows across the strike of the
rocks, and the gold is found below a hard bar; a, surface
of stream; b, mud and gravel forming bed of stream; c,

bedrock; *d,* auriferous gravel retained by the projection of the bedrock.

In Fig. 3 the stream flows as in Fig. 2, across the strike

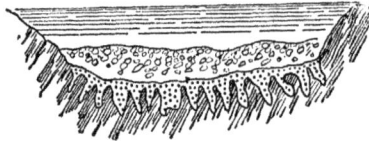

FIG. 3.

of the rocks, but the gold is found on one side of the creek: *a,* bank of stream; *b,* mud and other worthless

FIG. 4.

matter lying on the pay dirt; *c,* auriferous gravel accumulated in the deepest parts of the stream.

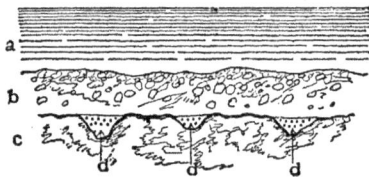

FIG. 5.

In Fig. 4 and 5, *a* represents the stream; *b,* mud and gravel at bottom of stream; *c,* bedrock; *d,* pot-holes in bedrock where auriferous material has lodged. In these the stream generally runs with the strike of the rocks, or at a slight angle; but the dip is nearly perpendicular in those instances where pot-holes have been found.

In estimating the value of alluvial claims it is of the utmost importance to consider the cheapness and abundance of the water supply, and, what is of no less importance, the facilities afforded by the surrounding levels for the disposal of the debris from mining operations.

Along the Pacific Coast — California, Oregon, British Columbia, and as far up as Alaska — are a number of auriferous deposits known as beach placers. They appear to be surface concentrations due to wave action, and an enrichment of this type seems to take place after heavy storms. Under such conditions the waves cut back into the coastal-plain sediments and concentrate the heavy material as a surface layer. The occurrence of these deposits is of interest, because they indicate that certain regions are auriferous. The richest deposits of this character are the famous beach placers at Nome, Alaska. A great deal of time and money has been wasted on beach sands at the other places mentioned, and the U. S. Bureau of Mines in 1919 reported that while there were spots worth investigation, the deposits were not worth further attention.

Indicative Plants. From early times it has been noted that the soil overlying mineral veins is covered by special vegetation, and though such cannot be taken as an infallible indication of the existence of mineralization, it will be interesting to record the results of past observations, so that they may serve as a guide to future observation:

The lead plant (*Amorpha canescens*) is said by prospectors in Missouri, Wisconsin, and Illinois to be most abundant in soil overlying the irregular deposits of galena in limestones. It is a shrub one to three feet high, covered with a hoary down. The light blue flowers are borne on long spikes, and the leaves are arranged in close pairs on stems, being almost devoid of foot-stalks. Gum trees, or trees with dead tops, also sumac and sassafras, are observed in Missouri to be abundant where 'float' galena is found in the clays.

A vein of iron ore near Siegen, Germany, can be traced for nearly two miles by birch trees growing on the outcrop, while the remainder of the country is covered with oak and beech.

The beech tree is almost invariably prevalent on limestone, and detached groups of these trees have led to discoveries of unsuspected beds of limestone.

Phosphate miners in Estremadura, Spain, find that the *Convolvulus althæoides,* a creeping plant with bell-shaped flowers, is a most reliable guide to the scattered and hidden deposits of phosphorite occurring along the contact of the Silurian shales and Devonian dolomite.

In Montana, experienced miners look for silver wherever the *Eriogonum ovalifolium* flourishes. This plant grows in low, dense bushes, its small leaves coated with thick, white down, and its rose-colored flowers being borne in clusters on long, smooth stems.

The 'zinc violet,' *Galmeiveilchen* or *Kelmesblume* (*Viola calaminaria*) of Rhenish-Prussia and contiguous parts of Belgium, is there considered an almost infallible guide to calamine deposits, though in other districts it grows where no zinc ore has been found. In the zinc districts its flowers are colored yellow, and zinc has been extracted from the plant. The same flower has been noted at zinc mines in Utah.

In Alaska, much of the gold-bearing gravel is frozen, necessitating thawing for its extraction. If a scrub willow is found growing on the surface, miners know that the ground is not frozen.

In the quicksilver-bearing serpentine belt of California brush and trees are sparse and stunted. This rock is pale blue and green. In other parts of this serpentine area where the reddish color of the soil indicates the presence of iron, the underbrush is fairly abundant. Some magnesite deposits are worked nearby, while the stream beds carry chrome float.

In examining a lode, the nature of the various minerals it contains and the proportions that these hold to each other should be observed. Sometimes it will be noticed that certain groups of minerals are often found together, the presence of one being favorable to the existence of the other. At other times the reverse will be noted, the existence of one mineral being the sign of the absence of another. The practical advantages to be derived from a series of observations indicating such results are too obvious to be overlooked. The following table, showing the association of ore in metalliferous veins, is given by Phillips and Von Cotta:

Two Members.	Three Members.	Four or More Members.
Galena, blende.	Galena, blende, iron pyrite (silver ores).	Galena, blende, iron pyrite, quartz *and* spathic iron, diallogite, brown spar, calc-spar, *or* heavy spar.
Iron pyrite, chalcopyrite.	Iron pyrite, chalcopyrite, quartz (copper ores).	Iron pyrite, chalcopyrite, galena, blende, *and* spathic iron, diallogite, brown spar, calc-spar; *or* heavy spar.
Gold, quartz.	Gold, quartz, iron pyrite.	Gold, quartz, iron pyrite, galena, blende; *and* spathic iron, diallogite, brown spar, calc-spar, *or* heavy spar.
Cobalt and nickel ores.	Cobalt and nickel ores, and iron pyrite.	Cobalt and nickel ores, iron pyrite; *and* galena, blende, quartz, spathic iron ore, diallogite, brown spar, calc-spar; *or* heavy spar.
Tin ore, wolfram.	Tin ore, wolfram, quartz.	Tin ore, wolfram, quartz, mica, tourmaline, topaz, etc.
Gold, tellurium.	Gold, tellurium, tetrahedrite (various telluride ores).	Gold, tellurium, tetrahedrite, quartz; *and* brown spar; *or* calc-spar.
Cinnabar, tetrahedrite.	Cinnabar, tetrahedrite, pyrite (various ores of quicksilver).	Cinnabar, tellurium, tetrahedrite, pyrite, quartz; *and* spathic iron, diallogite, brown spar, calc-spar; *or* heavy spar.
Magnetite, chlorite.	Magnetite, chlorite, garnet.	Magnetite, chlorite, garnet, pyroxene, hornblende, pyrite, etc.

CHAPTER II

THE BLOWPIPE

All chemical tests for minerals, whether made by the blowpipe or in the wet way, depend upon some chemical change, which allows the element, base, or acid, to be recognized. These changes consist either of decomposition of the mineral or the formation of new compounds. The following examples will illustrate sufficiently the nature of these changes:

If the oxide of a metal — copper for instance — is mixed with carbonate of soda and fused on charcoal, the copper is reduced to a metallic state, the oxygen combining with the charcoal to form carbon dioxide, which escapes as a gas, and any silica that is present decomposes the soda to form a silicate of soda, which is really a slag.

If a hydrous mineral is heated in a glass tube closed at one end, the water is given off, and condenses as drops in the cool part of the tube.

If an arsenic-bearing mineral is heated in a closed tube a crystalline deposit of arsenic is formed in the tube; but if it is heated in the air, white fumes of arsenious acid are evolved, which smell like garlic.

If a drop of hydrochloric acid be placed on a carbonate, such as limestone, the presence of carbon dioxide is recognized by the resulting effervescence; the stronger acid having combined with the lime liberated the carbonic acid in a gaseous form. In the case of many mineral carbonates, the acid requires to be heated for this reaction.

A great deal can be learned regarding a mineral by a few simple trials with the blowpipe, and every prospector should learn to use it. The chief requirements are a plain brass blowpipe 7 to 10 inches long, a candle, a pair of forceps or pliers, 6 inches of platinum wire, a small agate pestle and mortar, a small sieve, a magnet, a few small glass tubes, and some good hard charcoal free from cracks.

The only reagents that will be necessary are borax, carbonate of soda, and, rarely, microcosmic salt, nitrate of cobalt, and a little hydrochloric and sulphuric acid. A few others are occasionally needed, but their use is limited. The carbonate of soda should be perfectly dry, not merely dry to the touch, but quite free from water. Such soda may be prepared from common washing soda by expelling the water it contains. Put the soda in a shallow, clean, iron dish, and place it over a clear fire until a white, dry powder is formed, avoiding too strong a heat, otherwise the dry powder might fuse. A quarter of an ounce may be kept in a well-corked bottle or tube for use. Bicarbonate of soda may be used instead without previous heating; or if this be heated moderately it loses weight, and becomes carbonate of soda, free from water, like the other.

Borax should be dried in the same way; a quarter of an ounce will be enough. It is convenient to keep the platinum wire in the same tube. Unless these tubes are well corked, these chemicals re-absorb moisture. For testing tin ore it is advisable to have a little cyanide of potassium kept in a bottle, with the cork and rim well

covered with melted beeswax or paraffin; otherwise it
would absorb moisture, liquefy and be useless. It is a
deadly poison, and the greatest caution must be observed
in its use.

There are many types of good, cheap blowpipes.
The one shown in Fig. 6 has a tip that may be un-
screwed for clearing the hole, and a wooden mouthpiece.
The blowpipe should have a fine jet, or aperture, wide

FIG. 6.— A BLOWPIPE.

enough to admit a fine needle. The mode of using it
may be acquired readily by first breathing through the
nostrils with the lips closed, then puffing out the cheeks
(as if rinsing the mouth with water), still keeping the
lips closed, and breathing as before. The blowpipe may
at this point be slipped between the lips, and it will be
found that a current of air escapes through it without any
effort on the part of the operator. Air flows through the
pipe owing to the tendency of the distended cheeks to
collapse; it must never be forced from the lungs. After
a little practice the strength of the current may be increased. By breathing entirely through the nostrils, and
keeping the lips closed, the blast may be kept up for ten
minutes or longer without exhaustion or inconvenience,
except a slight fatigue of the lips in holding the blowpipe. The beginner may practice blowing upon a piece
of charcoal. The charcoal should, for convenience sake,

be cut into blocks 6 inches long by ¾ inches wide and ½ inch thick. Place a piece of lead, or a pin-head, or fragment of iron pyrite near the end of the charcoal, and learn to blow the flame of a candle to a point upon the object. However awkward the blowpipe may feel at first, practice will soon make the beginner expert. At first it may be necessary to gouge a small hole or recess

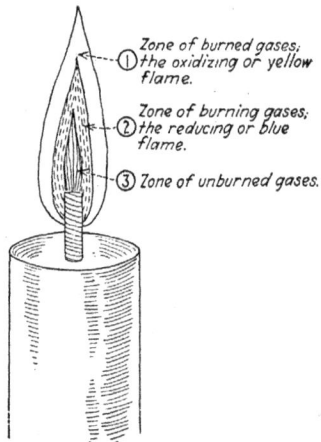

Zone of burned gases;
①the oxidizing or yellow
flame.

Zone of burning gases;
②the reducing or blue
flame.

③ Zone of unburned gases.

FIG. 7.—ANALYSIS OF A FLAME.

in the coal with the point of a knife, in order to prevent the specimen from being blown away; but after a while the blast may be used to burn out a cavity, although the depression cut by a knife or other instrument is best.

A study of the flame of a burning candle, gas jet, or spirit lamp is now necessary. Few people know or ever consider that there is anything to notice in a flame, but there is, as is shown in Fig. 7, and the zones are all important in blowpipe work. These act differently on the same substance. The outer zone is called the 'oxidizing flame,' and the inner, the 'reducing flame.' By blowing properly, these two flames may be made to turn horizontally, or even downward, and then either the outer flame or the inner flame may be turned on the 'assay,' as the object on the charcoal may be called. As a test, get

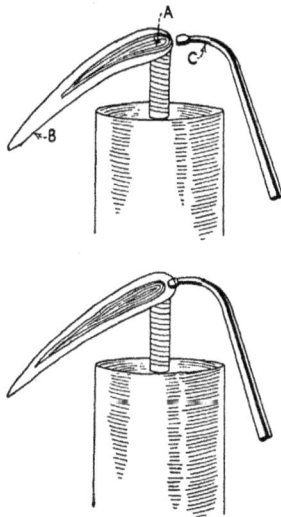

FIG. 8.— *A*, BLUE OR REDUCING FLAME; *B*, OXIDIZING FLAME; *C*, END OF BLOWPIPE.
By placing the end of blowpipe in the flame, the oxidizing flame *B*, is made more efficient.

a piece of iron ore as large as a pin-head, place it in a cavity in the charcoal, cover it with carbonate of soda as large as the assay, then turn the reducing flame down on the charge or mixture, when in a few seconds the ore will melt and be reduced to metallic iron. A magnetized knife-blade will pick up the prill or button of iron and the

soda. In this experiment a piece of red or brown hematite, or a piece of iron pyrite should be used, as neither
will be attracted by a magnet before the ore is reduced
to metallic iron. The reason for this action on the part
of the ore is that the mineral is iron combined with oxygen, and the reducing flame calls for more oxygen than
it possesses, so that when it is turned upon the hot oxide
of iron it takes the oxygen it calls for from the ore, and
leaves the iron in a metallic state. In the pyrite, which
is iron and· sulphur, the latter is partly driven off by
either flame. This process, on a larger scale, is called
' roasting.' The soda absorbs part of the sulphur and
part remains in the iron, but not so much but that the
magnetized knife-blade will attract it. The last experiment is good for practice, but not for illustrating the
two properties of the flame.

For proper blowpipe tests, all rocks should be ground
to fine powder, so that the borax bead on the platinum
wire may easily pick up the sample.

The following is a good illustration of the properties of
either zone of a flame. Wrap some platinum wire of the

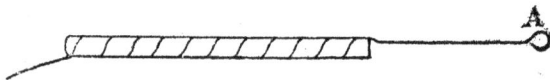

FIG. 9.— APPEARANCE AND SIZE OF WIRE AND OF
LOOP, *A*.

size of a large horse-hair around a match, leaving $1\frac{1}{2}$
inches extending beyond the match end, and roll the end
of the wire around another match until a small loop is
made (Fig. 9). Prepare a little borax powder, and after
heating the loop in the main flame, plunge it quickly into
the powdered borax. It will immediately pick up some

of the powder, and, by turning the flame, on the borax, a clear and transparent bead will fill the little loop at the end of the wire. The apparatus is now ready to show the use of the oxidizing and reducing flames.

Heat the borax bead red-hot in the main flame, and sprinkle on it or dip it into a little black oxide of manganese, which will stick to the bead. Then turn the outer or oxidizing flame on the bead, and blow until the manganese has been entirely dissolved or melted — it will impart to the bead a beautiful amethystine-purple. Next turn the inner or reducing flame on the bead, and in a few seconds the color will disappear, but will return when the outside flame is used again.

If the platinum loop will not hold the borax bead, it is too large. If it is dimmed or blackened by smoke, heat it red-hot — it will clear up.

The reactions of different substances are given throughout this book as they are called for when they are described.

A glass tube of a little less than ⅜ inch in diameter may be made into a blowpipe as follows: Take a piece of tube, 10 or 12 inches long, soften to red heat in the flame of a spirit lamp, and draw it out to a thin point. Cool and break it off at the smallest diameter, so that air may be blown through. Put the tube into the flame again and bend the glass to a right angle, about two inches off from the point. Cool gradually, and heat the mouth end, opening it a little by introducing a small, dry, pine stick, when finally one has an efficient blowpipe when one of metal is not available.

The three principal means of chemically testing minerals before the blowpipe are (1) with borax; (2) on

-charcoal, usually with the addition of carbonate of soda; (3) by holding in the oxidizing point. If, in a test the borax bead has too deep a color, or is opaque, it shows that too much powdered mineral has been used; all beads should be transparent in order to judge the correct coloration.

The following method may be used to show the presence of certain metals:

COLOR OF BEAD IN

Oxidizing flame	Reducing flame	Indicates presence of
Green (hot); blue (cold)........Red ....Copper.		
Blue (hot and cold).............Blue ....Cobalt.		
Amethyst ....Colorless ....Manganese.		
Green ....Green ....Chromium.		
Red or yellow (hot) ....⎫ Yellow or colorless (cold) .....⎭Bottle-green ......Iron.		
Violet (hot); red-brown (cold). Gray and turbid, dif- ficult to obtain..Nickel.		

It requires some practice before reliable results can be obtained in reducing. The reduced bead, if brought out of the flame at a white heat into the air, may at once oxidize; but this may be prevented by placing it inside the dark inner cone of an ordinary candle flame, and allowing it to cool partly there.

The blowpipe, soda, and charcoal method is designed to extract metals from minerals; it favors in the highest degree the removal of oxygen. But, like the borax test, it is limited in its application, as it can only be used to detect certain metals. The failure of the test in any case must not be looked upon as conclusive proof of the absence of the particular metal sought; for instance, copper can be extracted easily from its carbonate by this test, but not from copper pyrite. Still the test is a valuable and indispensable one. The test is complete when the metal is obtained as a globule in the cavity of the charcoal. In many cases the globule will be found surrounded by the oxide of the metal, forming an incrustation on the charcoal; and the color of such should be carefully noted, both at the moment of removal from the flame and after cooling. By pressing the globule it can be determined whether the metal is malleable or brittle. The following observations and inferences may result from this test:

Globule	Incrustation	Presence of
Yellow, malleable ...None		Gold.
White, malleable ....None		Silver.
Red, malleable ......None		Copper.
White, malleable ....White		Tin.
White, malleable ....Red (hot); yellow (cold)		Lead.
White, brittle ....Red (hot); yellow (cold)		Bismuth.
None ....Yellow (hot); white (cold)		Zinc.
White, brittle, giving off fumes when removed from flame.White		Antimony.

The blowpipe is useful in the detection of sulphur and arsenic. This may be done by heating a piece of the mineral, supported on charcoal or held in a forceps in the oxidizing point of the flame, and comparing the odor given off. The smell of burning sulphur is well known, while white fumes having a garlic odor indicate the presence of arsenic.

Mercury, antimony, and other substances may escape as fumes when heated in this manner.

Nitrate of cobalt dissolved in water, used in very small quantity, helps to discriminate between certain white

minerals, such a kaolin, meerschaum, magnesite, dolomite, etc. The mineral is reduced to powder and moistened with a drop of a very light solution, and then heated
before the oxidizing flame of the blowpipe. Kaolin and
other minerals containing alumina assume a rich blue
color, while meerschaum and others containing magnesia
become flesh-colored. Oxide of zinc, under the same circumstances, becomes green. This may be tried with the
white coating obtained on charcoal by reducing an ore
of zinc with carbonate of soda.

Tests in glass tubes can be made better over a spirit
lamp, so as to avoid the deposit of soot on the glass; but
they may also be made with the blowpipe flame, provided
that it is used carefully, avoiding too sudden a heat,
which would break or fuse the glass. The presence of
water in minerals is detected in this way, as the water condenses and collects as small drops in the cold part of the
tube. Some minerals containing sulphur, arsenic, antimony, tellurium, and selenium often give a characteristic
deposit.

Minerals containing mercury can also be tested in this
way, as by adding a little soda, sometimes with cyanide of
potassium, a sublimate of metallic mercury will be formed
in the cold part of the tube. A little charcoal should be
added to arsenical minerals.

Organic combustible minerals generally leave a deposit
of carbonaceous matter at the bottom of the tube, and
the volatile hydrocarbons condense in the cooler part.
The tube should, therefore, always be long enough to
allow for this condensation. Minerals that yield a characteristic smell will be best tested in this way.

As a good blowpipe set is very important in prospecting, the following list and illustration prepared by the Braun-Knecht-Heimann Co. of San Francisco is worth study:

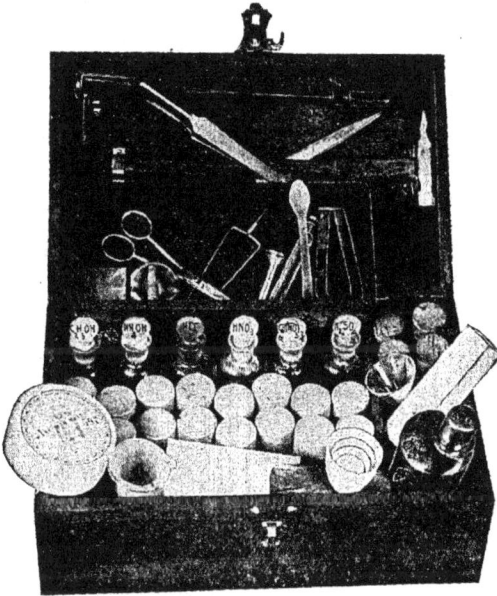

FIG. 10.

This set is packed in varnished wood box, 12 by 7 by 5 inches, outside measurements. It contains forty-eight different articles of apparatus and chemicals packed in bottles, boxes, and recesses, neatly labeled and ready for use, and costs $20.

1. Anvil, polished steel	6. Clay cylinder
2. Brass blowpipe	7. Charcoal stick
3. Mortar	8. Evaporating dish
4. Alcohol lamp	9. 3 porcelain crucibles
5. Beaker	10. Forceps, 5-inch

11. Forceps, palau tips
12. Hammer
13. Horseshoe magnet
14. Pipette
15. Platinum wire holder
16. 3 pieces platinum wire
17. Scissors
18. Horn spoon
19. Test lead measure
20. 3 open tubes
21. 3 closed tubes
22. Magnifier, 1-inch
23. Blue litmus paper
24. Red litmus paper
25. Turmeric paper
26. Brazil wood paper
27. Borax glass
28. Borax powder
29. Bone-ash

30. Antimony oxide
31. Arsenous acid
32. Salt phosphorus
33. Soda carbonate
34. Soda nitrate
35. Alcohol
36. Ammonia hydroxide
37. Hydrochloric acid
38. Copper oxide
39. Iron oxide
40. Lead oxide
41. Manganese oxide
42. Mercury oxide
43. Potash bisulphate
44. Test lead
45. Tin oxide
46. Cobalt nitrate solution
47. Nitric acid
48. Sulphuric acid

Another set, containing everything as above, but measuring approximately 12 by 7 by 6½ inches, with extra tray containing the following minerals in screw-cap vials neatly labeled; costs $25.

1. Graphite
2. Sulphur
3. Stibnite
4. Chalcopyrite
5. Galenite
6. Rutile
7. Pyrrhotite
8. Pyrite
9. Hematite
10. Magnetite
11. Chromite
12. Limonite
13. Siderite
14. Pyrolusite

15. Rhodonite
16. Sphalerite
17. Willemite
18. Corundum
19. Cryolite
20. Fluorite
21. Calcite
22. Apatite
23. Gypsum.
24. Dolomite
25. Barite
26. Magnesite
27. Celestite
28. Strontianite

29. Halite	40. Lepidolite
30. Quartz	41. Chrysolite
31. Orthoclase	42. Scapolite
32. Witherite	43. Tourmaline
33. Albite	44. Cyanite
34. Spodumene	45. Pyrophyllite
35. Hornblende	46. Talc
36. Wollastonite	47. Datolite
37. Beryl	48. Prehnite
38. Garnet	49. Pectolite
39. Mica	50. Stilbite

The following is a prospector's blowpipe set, as arranged by Eimer & Amend of New York. It is fitted in a wooden case and costs approximately $10. The set contains:

1 jeweler's blowpipe.
1 alcohol lamp.
1 magnifying lens, double
1 porcelain mortar, 2¼ in.
2 porcelain crucibles and covers.
1 funnel glass, 2 in.
1 dozen test-tubes, 3 in.
1 dozen glass tubes and rods, assorted.
3 small beakers, 0 to 000.
1 pair slag pincers.
1 spatula, 3 in.
1 piece sheet zinc.
1 piece copper wire.
1 piece tin foil.
1 chamois skin.
1 H. S. magnet, 3 in.

1 piece iron wire, platinum wire and holder.
3 carbon sticks.
1 package filter-paper.
2 drams ferrous sulphate.
2 drams borax glass.
2 drams oxalic acid.
2 drams sodium carbonate, dry.
1 oz. sulphuric acid, commercially pure concentrated.
1 oz. muriatic acid, c. p. conc.
¼ lb. nitric acid, c. p. conc.
¼ lb. ammonia, strong.
4 oz. alcohol.
2 oz. mercury.
2 oz. granulated lead.
2 drams carbonate potash.

In addition to these the outfit should include a small hammer with ½-in. square face, a pair of forceps and pliers, and plaster tablets, which are used for the same purpose as the charcoal sticks.

CHAPTER III

CRYSTALLOGRAPHY

The forms that many minerals take always indicate their composition. It is, therefore, sometimes a great help to the prospector if he understands the principles of crystallography, so far as to enable him to determine the system or order to which a crystal belongs. Nearly all minerals, when they appear in a crystalline condition, assume individual characteristic forms. Although, to the unaided eye and unskilled vision, this assertion may appear to be incorrect in a few instances, it appears so only because the differences are exceedingly small.

All crystalline forms have been reduced to six classes or systems, which are: (I) isometric; (II) tetragonal; (III) hexagonal; (IV) orthorhombic; (V) monoclinic; and (VI) triclinic.

Isometric system. The principal forms of this system are the cube, octahedron, dodecahedron, the two trisoctahedrons, the tetrahexahedron, and the hexoctahedron.

A cube has six equal and square sides, as in Fig. 11. In this form, lines drawn from the center of each face to the face opposite, cross each other at right angles, and are of the same length.

This system is called ' isometric,' from *iso,* equal, and *metric,* measure, because these axes or lines are of equal

length and at right angles to one other. It must, however, be remembered that the cube is modified in some
minerals, but wherever these modifications take place
the original form of the cube may always be traced.
Some of the changes may be intricate, but these unusual
forms will not be discussed here; the common forms
only are of importance.

The student should take a potato, cut as perfect a cube
as possible, and familiarize himself with the common
variations that may belong to the cube, without changing
the length of the axis, and cutting so that the axis will
always be the same or of equal lengths.

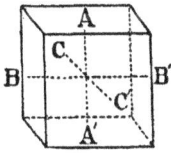

FIG. 11.— THE   FIG. 12.— THE   FIG. 13.— THE
    CUBE.   OCTAHEDRON.   DODECAHEDRON.

Fig. 11 is the cube with the three axes $A A'$, $B B'$, $C C'$.
If, with a knife, one edge angle from A to C' and from
A to C is sliced off, and similarly from A to B' and from
A to B, a four-sided pyramid is the result, the apex of
which will be at A and the four-sided base at $C B'$, $C' B$,
or around one-half the cube. Treat the opposite side in
the same way, and we have an octahedron (Fig. 12).

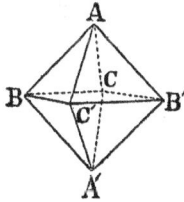

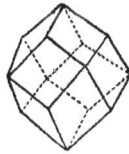

The dodecahedron (12 sides), Fig. 13, may be formed
by taking off the solid angles A, B, B', A'. In all three
cases, and in many others, the three axes remain the same

in length and in their angular direction, where the forms have not been distorted.

Tetragonal system. The principal forms of this are the two square prisms and pyramids, and the eight-sided prism and double eight-sided pyramid.

The tetragonal system has also three axes, as in the isometric; they are at right angles to one other, but the vertical axis is longer than the others, as in Fig. 14.

' Tetragonal ' means four-cornered or angled. This is not precise, as a cube is tetragonal; yet it is used to ex-

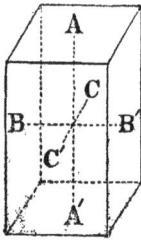

FIG. 14.— TETRAGONAL   FIG. 15.— TETRAGONAL   FIG. 16.— THE
PRISM.   OCTAHEDRON.   ZIRCON.

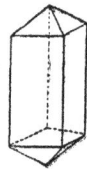

press this form because it is one word, otherwise ' square prismatic ' would be a more correct term, since Fig. 14 is that of a prism. In mineralogy, any crystal having parallelograms for sides, is called a prism. Cut this prism as in the case of the cube, and we have the form shown in Fig. 15.

Variations in this form may show a prism with four-sided termination at either or both ends, as in Fig. 16. This is the form of the transparent gem called the zircon, once called the jacinth. The zircon has been mistaken for the diamond, which it resembles in brilliancy, but its hardness is 7.5 compared with 10 for the diamond and specific gravity 4.7 against 3.5. Zircon is formed in granites and syenites. The diamond is isometric and never tetragonal, hence it may be distinguished readily from the zircon.

Hexagonal system. The chief forms of this system are the two six-sided prisms, the two double six-sided pyramids, the twelve-sided prism and double twelve-sided pyramid. It differs from the tetragonal system in that it has three equal lateral axes instead of two, the vertical

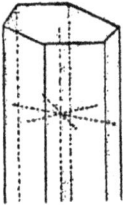

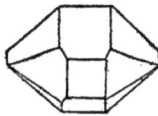

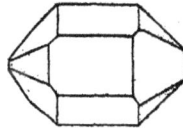

FIG. 17.— HEXAGONAL   FIG. 18 and 19.— QUARTZ CRYSTALS —
PRISM.   HEXAGONAL

being at right angles, as in Fig. 17, with each of the three lateral.

It must be remembered that owing to various causes the hexagonal crystal always calls for hexagonal terminations; see Figs. 18 and 19.

This crystal may be found under various modifications of the hexagonal form, but it can always be reduced to this system. The symmetry of the crystals may be by sixes, or rarely, by cutting each angle; it may be in twelves, or the sides may be unequal in area or length, as in Fig. 18. A quartz crystal from Switzerland was three-sided for nearly its entire length, but showed its hexagonal nature only at the extremity, where, having been free from its confinement in process of formation, it had assumed its normal crystallization. Calcite crystals some-

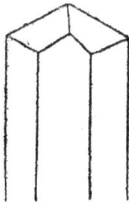

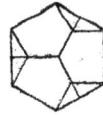

FIG. 20.— CAL-
CITE HEXAGONAL
CRYSTAL — THREE-
SIDED TERMINATION.
SIDE VIEW.

FIG. 21.— THE SAME
— END VIEW.

times assume a hexagonal prism precisely as does quartz, but the latter always shows six-sided terminations,

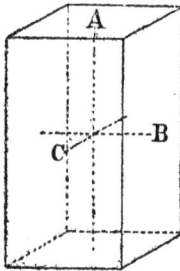

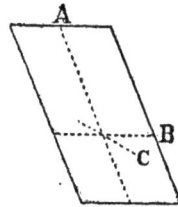

FIG. 22.

FIG. 23.

whereas lime or calcite crystals show three-sided terminations, as in Fig. 20 and 21. There are two sections or forms of this system — the hexagonal and the rombohedral — both belonging to the hexagonal system, and identified as shown.

These calcite crystals belong to the rhombohedral section of the hexagonal system, showing rhombohedral forms at the end, as in Fig. 15.

Orthorhombic system. The characteristic forms of this system are the rhombic prism and pyramid. There are also other forms called domes. In this system the three axes are unequal and intersect at right angles, as in Fig. 22, wherein the axes, $A, B, C,$ are unequal in length, but at right angles at the intersection. The terminations are flat, although frequently beveled on the surrounding edges.

Monoclinic sytem. These forms are too complex to be fully described here, but it is not difficult to learn their essentials. In this system, two of the axial intersections are at right angles but one is oblique, and the side of the crystal is inclined, as in Fig. 23.

Generally, crystals of feldspar, which contain potash (called orthoclase or potash feldspar), are monoclinic; but the soda feldspar crystals belong to the next or sixth system, as do also the lime feldspars.

Triclinic or *'thrice inclined'* system. In this the planes are referred to three unequal axes all oblique to one another. The only important feature in the system is that there is no right angle in any of its crystals; but it is of little use for ordinary study, since with the exception of the lime-feldspar and soda-lime feldspars (anorthite or lime-feldspar, labradorite or soda-lime feldspar; andesite and oligoclase, both soda-lime feldspars; and albite, a soda feldspar) all the others are of little importance, except microcline, a potash feldspar.

The following are examples of the foregoing:

Isometric — gold, silver, platinum, amalgam, copper, diamond, garnet, magnetite, pyrite, galena, alum, kalinite, all of which assume the cubic octahedral, or more allied form.

Tetragonal — zircon, chalcopyrite, cassiterite (tin ore), titanic oxide, and others.

Hexagonal — beryl, aquamarine, emerald, chrysoberyl, apatite (lime-phosphate), quartz.

Orthorhombic — barite or sulphate of barytes, celestite or sulphate of strontia, and carbonate of strontia, also cerussite or lead carbonate.

Monoclinic — borax, gypsum, glauber salt (mirabilite is its mineralogical name), copperas (or melanterite).

Of the sixth system, sufficient examples have been given.

Of the gems not mentioned above, the turquoise owes its blue to copper, and is never crystallized, being in reniform or stalactitic conditions. It is a phosphate of alumina with water in combination. This mineral or gem should be carefully distinguished from lazulite, which, though blue, crystallizes in the monoclinic, or fifth system. It is a softer mineral, and is a phosphate of alumina, magnesia, and iron. Lazulite is found in beautiful crystals at Crowder's Mount, in Lincoln county, N. C.; also 50 miles north of Augusta, at Graves's Mount, in Lincoln county, Georgia; also in California, but it is a rare mineral.

CHAPTER IV

SURVEYING

There are a few simple measurements that are sometimes needed, which can be made without carrying instruments. The actual work of surveying, to be of any value to the prospector, must be so accurately performed that the subject should be treated as a specialty, whereby a theodolite or transit and logarithm tables must be used. Any elementary book on surveying or trigonometry will give sufficient information.*

A few measurements, and simple surveys with easy methods, are given here to meet cases where only a general approximation is required.

TO MEASURE HEIGHTS THAT ARE INACCESSIBLE

The height of a tank, chimney, stand-pipe, tree, or other object, may be measured approximately by knowing one's own height and taking advantage of sunlight, for example, consider Fig. 24:

Let *A B* be the height of the object to be measured; the dotted line is the shadow cast. Walk into the sunlight, and mark on the ground the point at which your own shadow terminates; measure from the heel to that

---

* For this purpose we would recommend the following: " The Practical Surveyor's Guide." By Andrew Duncan. New York, Henry Carey Baird & Co., Inc. Price, $1.75.

point. A calculation in single rule of three will give *A B* thus:

$$C' B' : B' A' :: B C : A B$$

Heights of hills or land surfaces may be measured near enough by the aneroid barometer. Instructions for its

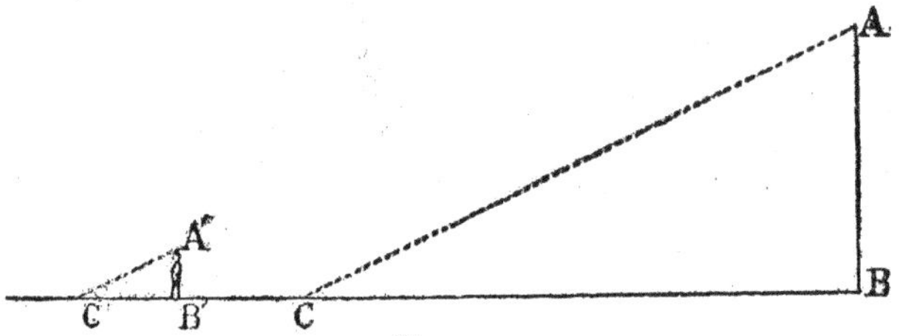

Fig. 24.

use are given with the instrument. Fairly accurate aneroids may be purchased small enough to go into the side pocket; or still more accurate ones may be carried easily in a case held around the shoulders by a small strap.

The aneroid barometer is a very useful instrument that should be included in a field outfit. It shows the altitude or elevation above sea-level, being sensitive to variations in atmospheric pressure. The aneroid is not as accurate as the mercurial barometer, but has the advantage of being portable and giving approximate readings. It is made in watch form 1¾ inches diameter, as a small clock 2¾ inches diameter and 1¼ inches deep, or in a brass case 5 inches diameter, and is good for elevations of 3000 to 8000 feet, marked in 50-foot divisions. The instrument is easy to read, so the prospector can plot

FIG. 25.— AN ALTITUDE (ANEROID)
BAROMETER.

An example of the use of this instrument, which is the Tycos
type made by the Taylor Instrument Companies of Rochester,
New York, and costs $35 with leather case, is as follows:

Suppose that a person is going from Spokane, Washington, to
Boise, Idaho: rotate the altitude scale until the hand points to
1,900 feet, the elevation of Spokane, and during the 350-mile
journey the hand will reach the 2,500-foot point, the altitude of
Boise. Another method is to set the ' o ' of the altitude scale
opposite the point of the hand. so that if a prospector is going
across country from Eureka, California, to Red Bluff, in the same
State, he would cross several high ranges — up to several thousand feet — and arrive at his destination with the aneroid reading
300 feet, the instrument showing all the variations in height during
the trip. The inner dial is divided into inches of pressure or into
barometric pressure so that this part can be used for the purpose
of weather forecasting.

roughly the ground over which he is traveling, or the fall in streams, and such like.

TO MEASURE AREAS

Theoretically, it is easy to ' step-off lines,' but practically it is difficult by this method to be correct on uneven land. Where one is acquainted with the exact average measurement of his step on level land, he may reach some approximate accuracy on uneven land by remembering that in ascending, even slightly, his average decreases, and the reverse in descending. A good strong tape-measure, kept on a level in ascending and descending hills, is more convenient and more easily handled than a chain.

On square areas, the length of the side multiplied by that of the adjacent side, gives the area.

In the parallelogram, where all angles are right angles, the same rule applies.

In any other shape the following rules should be observed: First: measure the area of a right-angled triangle thus:

Let B, Fig 26, be the right angle; the area of $A B C$ is equal to the length, $B C$, multiplied by half the perpendicular distance, $A B$.

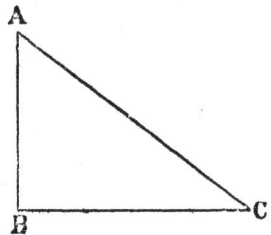

FIG. 26.

Example: $B C = 100$ ft.; therefore, if $A B = 90$ ft., $100 \times 45 = 4500$ sq. ft. $=$ area of $A B C$.

The same rule applies when the triangle is not a right-angle triangle; thus, the angle at A, Fig. 27, is obtuse.

$D C = 150$ ft., $A B = 90$ ft.; multiply 150 ft. by one-half *A B* or 45 ft., giving 6750 sq. ft., as *A C D* is composed of two right-angle triangles, *A C B* and *A B D,* as in the previous example.

When the triangle has an acute angle at *A,* as in Fig. 28, treat precisely as in Fig. 27, only letting the perpen-

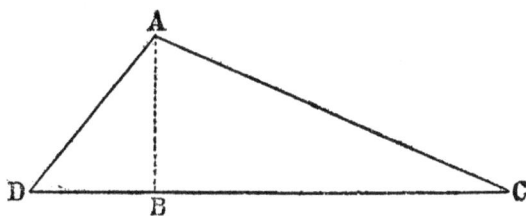

FIG. 27.

dicular fall from *D* upon *A C,* that is, invert the triangle.

The cases wherein the sides are more than three are treated by resolving all such areas into right-angle triangles, thus:

In Fig. 29 the area, *A C D B,* may be resolved into two

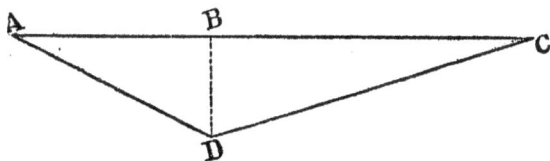

FIG. 28.

triangles, *A C B* and *C D B,* of which *A B* is the base of the one and *C B* that of the other. In Fig. 30, the area, *A C D B E K,* may be resolved into the four triangles, *A C D, A D B, A B E* and *A E K.* The perpendiculars of Fig. 29 are *E D* and *C F.* Those of Fig. 30 are *C H*

I B, F E and *K G,* and the length of bases may be multiplied by half that of the perpendiculars, as in the case al-

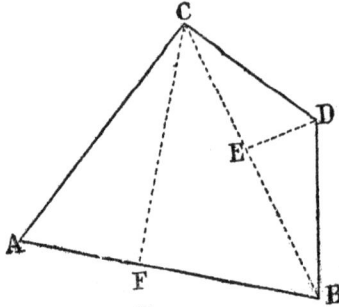

FIG. 29.

ready given, and the feet be reduced to acres, rods, or miles.

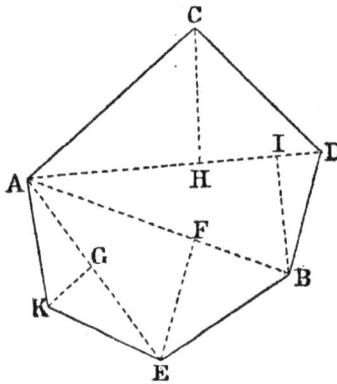

FIG. 30.

Suppose the area of Fig. 30 be 80,000 sq. ft., then according to table No. 3 in the Appendix it will be 1 acre, 3 roods, 13 poles, 25 yards, 7 feet, or 1.836 acres. (There are 43,560 sq. feet in an acre.)

TO MEASURE AN INACCESSIBLE LINE

Suppose it is desirable to measure the distance — *A B* — across a river, as in Fig. 31.

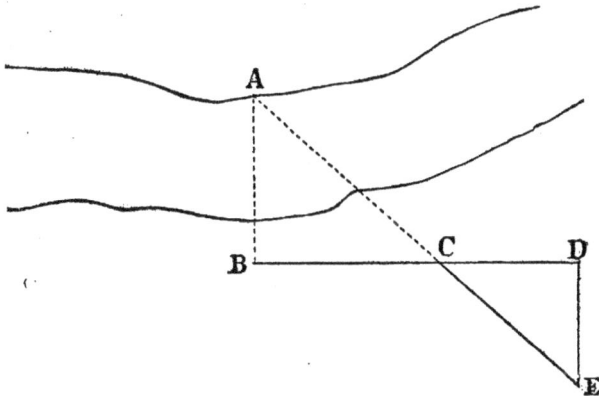

FIG. 31.

Measure a distance of about 100 ft., *B D*, at right angles to *A B*, and raise a pole at *C*, about half-way from *B* to *D*. Proceed in measuring at right angles to *B D*, in the direction *D E*, letting *E* be that point at which the line *C E*, if extended, would strike *A*. Now there are two right-angle triangles of the same angles, for, as every triangle has two right-angles according to geometry, and each of these triangles has one right angle, and the opposite angles at *C* are equal according to geometry, the remaining angles at *A* and *E* are equal, and the triangles are proportional, and the proportion is

$$C\,D\,:\,D\,E\,::\,C\,B\,:\,A\,B.$$

Then, if $C\,D = 40$ ft., $D\,E = 45$ ft., and $C\,B = 60$ ft., therefore $45 \times 60 = 2700$ divided by $(C\,D)$ 40 ft. $= 67\frac{1}{2}$ ft.; this is for $A\,B$, or the distance across the river.

The only difficulty is in measuring the angles as true right angles, and this may be done by measuring the perpendicular, thus:

Extend the line $A\,B$ (Fig. 31) to F (Fig. 32) and likewise the line $D\,E$ (Fig. 31), to C, as in Fig. 32. Then measure equal distances on the line $B\,D$, for the lines or

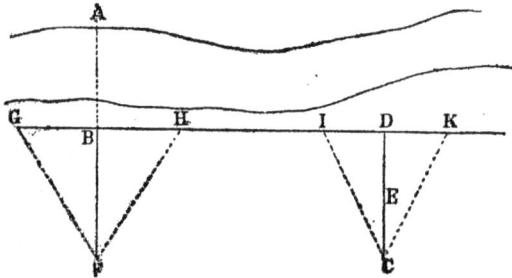

FIG. 32.

offsets, $B\,C$ and $B\,H$; also from $D\,C$, the offsets $D\,I$ and $D\,K$; drive stakes in at G, H, I, and K. See that the distances represented by the dotted lines are equal, and if so, the lines $A\,B\,F$ and $D\,C$ are perpendicular to the line $G\,K$, and the result will be nearly accurate.

It is well for the prospector to use a prism compass, which will read to one-quarter degree. Such an instrument may be bought at reasonable cost, not more than 3 in. diameter, of light weight and of sufficient accuracy. The Brunton is an excellent transit for general purposes, and is used extensively.

In almost every surveying project, especially in driving adits and sinking shafts to connect with tunnels and drifts, only the best instruments should be used. Everything depends upon accurate measurements, and this phase of engineering is not one that can be treated approximately, because any error may result in provoking and expensive

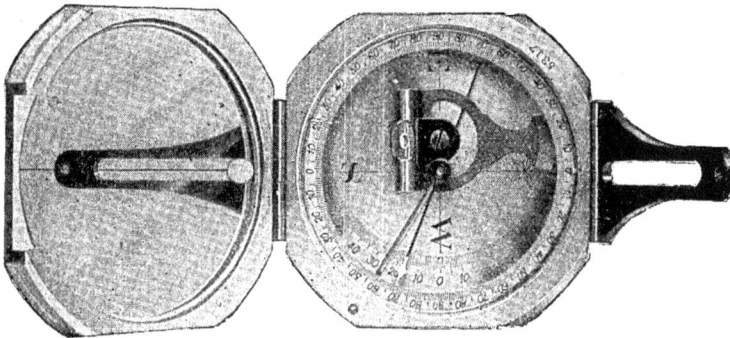

FIG. 33.

The Brunton pocket transit, is an instrument, which performs, within the limits of accuracy imposed by its size and construction, the operation for which the ordinary transit is used. It is the lightest and most convenient pocket transit on the market. No tripod or Jacob-staff is necessary, as the sighting and reading are done simultaneously. The case is of aluminum alloy 2¾ by 2¾ by 1⅛ inches, and completely encloses the instrument. The magnetic variation is set off by means of the slotted head pinion placed in one corner of the case, which may be revolved by means of a screw-driver or knife-blade. While the transit is designed to be used principally as a hand instrument, it is sometimes desirable to utilize the advantages of a fixed support, and an attachment is furnished which permits the operator to use for this purpose the light telescoping camera tripod that most engineers include in their traveling equipment. This instrument costs $25, plus $7.35 for tubular extension tripod, if desired.

mistakes. At the same time, the reviser of this edition has seen several remarkable surveys made with a tape-measure and compass, also good connections made. Of course these were in small mines, but nevertheless served the purpose.

A ready-reading tape is an improvement over the old type. The foot-numbers, which are repeated at every sub-number, are placed at right angles to the sub-numbers and are read across the tape instead of lengthwise. This arrangement facilitates reading and thus prevents errors and saves time. In making horizontal measurements greater than five feet, the tape-user is behind his tape, so that this lateral position of the foot-numbers is the most convenient, for both horizontal and vertical measuring. It is less confusing than where all numbers (foot and inch or tenth alike) are positioned longitudinally on the tape; in which case, foot-numbers and sub-numbers, being often duplicated, are frequently mistaken for each other. The foot-number is repeated at every inch-mark or tenth mark, directly ahead of the sub-number, throughout the entire length of the tape. This absolutely prevents mistakes in reading the tape, since there can never be the slightest doubt as to the number of feet measured at any point on the tape. This tape is made by the Keuffel & Esser Co. of New York, who also make the ' Favorite " farm level, which is a $22.50 instrument that is useful for measuring and laying-out tracts of land, or running levels, and such like.

CHAPTER V

ANALYSIS OF ORES

In the determination of ores, two kinds of analyses are employed, namely: the wet method, by the agency of chemicals, and the dry method, by heat and fusion with fluxes.

Preliminary examinations may be made with a pocket lens and a piece of steel or a heavy-blade pocket-knife, the first to see if any native metals or sulphides are present, and the second to try the softness or silicious nature of the mineral. If much quartz is present it will strike fire.

Pulverize an average sample of the ore and use the blowpipe to detect sulphur, arsenic, selenium by the smell, on charcoal, or in the glass tube. Arsenical fumes have a garlic odor; selenium that of horse-radish.

Use a test-tube with a little nitric acid, and heat over a spirit flame. Add a few drops of water and one drop of sulpho-cyanide of potash — an intense deep red appears, deeper according to the amount of iron and solubility of the mineral in nitric acid.

Try another portion in the same way, but add one drop of hydrochloric acid. A dense, curdy, white precipitate indicates silver.

Native gold or silver is determined by color and softness, as described elsewhere. Treat another portion in the same way with nitric acid, add several drops of strong liquid ammonia; an azure-blue color indicates copper. Antimony and tin are detected by the blowpipe. Place the former upon charcoal with carbonate of soda, and brilliant metallic globules are obtained; the metal and fume volatilize and cover the charcoal with white incrustations, and needle-shaped crystals appear. Tin is reduced when the ore is mixed with carbonate of soda and cyanide of potassium on charcoal, and the reducing flame turned on, ductile grains of metallic tin and no incrustations appearing.

Manganese gives amethystine beads of borax in the oxidizing flame, disappears with the inner reducing flame, re-appears with the oxidizing flame.

Alumina, manganese, lime, give their characteristic colors, or, with the last-named, incandescent light before the blowpipe on charcoal. Alumina heated on charcoal, and then touched by a half drop of proto-nitrate of cobalt, then heated strongly in the oxidizing flame, gives a blue color. Magnesia so treated gives a faint red or pink, seen just as it cools.

Zinc heated on charcoal with carbonate of soda in the reducing flame becomes metallic, and when oxidizing in the outer flame gives a white oxide which is yellow when hot, white when cooled; and with proto-nitrate of cobalt when heated in the outer flame a beautiful characteristic green color.

Cobalt and nickel give the colors that have been noticed on another page under their respective names.

Nickel, when present in ores, may be detected by the blowpipe; but the method is not very satisfactory when the amount of nickel is small, as the large quantity of other elements that may be present are likely to obscure the nickel reaction. The most sensitive test is that with dimethylglyoxime. If a little of this reagent be added to a solution containing nickel, then ammonia added to a slightly alkaline reaction, and the solution boiled, a red crystalline nickel salt of dimethylglyoxime is formed. If much cobalt is present, hydrogen peroxide must be added after the ammonia, and the excess boiled out before adding the dimethylglyoxime.

Uranium heated with microcosmic salt (phosphate of soda and ammonia) on platinum wire in the outer flame dissolves, producing a clear yellow glass, which on cooling becomes yellowish-green; in the reducing flame yellowish-green when hot, and green when cold. But the analyst should remember that copper also produces a green bead, but only in the outer or oxidizing flame, and chromium the same, but in both outer and inner flames.

The copper-green becomes blue on cooling, the chrome-green remains green on cooling. This will always distinguish between them.

Titanium, in the presence of peroxide of iron, as in some titanic ores of iron and sand, gives, with microcosmic salt in a strong reducing blowpipe flame, a yellow glass, which on cooling fades out to a delicate violet.

Vanadium, dark yellow when hot, paler on cooling; in the reducing flame brownish-red when hot, chrome-green on cooling.

Mercury may be detected in almost any of its ores by heating in a glass tube and noting, under the lens, its sublimation as minute shining particles.

Minerals that are carbonates may be detected by their effervescence when touched by a drop of hydrochloric acid, as in limestone and spathic iron ore. But the analyst must remember that some cyanides effervesce where neither lime nor carbon dioxide is present, and chloride of lime where there is no carbon dioxide. With these latter other tests must be used, but the sense of smell will show that carbonic acid does not exist, the latter having no odor.

Some sandstones contain a small quantity of lime carbonate and must be tried under the lens, as the bubbles are minute. While great help is received from these tests, and many determinations can be made, especially in simple minerals and ores, there are ores so complex that the above tests fail to give satisfaction, because the colors are mixed and the reactions are confused. Some of the elements must be removed from their association and a separation made. This analysis is called 'qualitative,' and a complete analysis of a compound ore follows:

I. WET METHOD

There are many times when it becomes not only a matter of curiosity but of importance for the prospector to know the entire composition of an ore. With a little practice the 'wet method,' as it is called, may be used with all the accuracy required under the circumstances.

Although the dry method is simpler than the wet method for one or two elements such as gold and silver, it may so happen that sufficient heat is not procurable. Some directions for the wet method include:

1. Pulverize an average sample of the ore as fine as possible and pass the lot taken as an assay through a sieve of 80 meshes to the lineal inch, equal to 6400 holes per sq. in., being careful that nothing is left on the sieve.

2. Drop a little of the sifted ore into a test-tube, pour a little nitric acid upon it, add one-eighth part water, and warm gently over the flame of a spirit lamp to see if it will dissolve; if not, then add four times as much in bulk of muriatic acid (hydrochloric acid). If this will not dissolve, then proceed as follows:

3. Put the assay, after fine pulverization, into a platinum crucible. Place it in a suitably arranged platinum wire triangle so that it will hang over an alcoholic blast-lamp. When all is ready add a mixture of equal parts of sodium carbonate and of potassium carbonate, amounting in all to about four times the bulk of the assay, stir gently with a glass rod or a stiff platinum wire, and then light the lamp. Watch the assay, and when it begins to swell withdraw the lamp, but return it when this action subsides, so that the alkalies do not throw the charge out of the crucible, which should be only half full at the beginning. With care, the contents will soon subside, and under increased heat become a quiet, liquid mass. Extinguish the flame, cool the crucible, remove contents to a glass beaker or place the crucible, with its contents, within the beaker, and pour a little water upon it, add some nitric, or a little hydrochloric, but not the two acids together, unless only the assay, and not the platinum crucible, is in the beaker (nitro-muriatic acid or aqua regia dissolves platinum). Warm and stir till the assay is entirely dissolved, except perhaps some white grains of silica.

4. If the preceding work has been properly performed, the fused mass is now dissolved. Filter the contents of the beaker to separate any undissolved remainder, if any such is seen in the glass, and wash the filter-paper by passing an ounce or two of water through it. It is not neces-

sary, where extreme accuracy is not required, to wash the filter-paper perfectly free from the acids; but if it be necessary, get a small strip of platinum ribbon; and polish its surface. If a drop of the filtrate evaporated from this surface shows no trace of sediment or outline, even under a lens, the filter-paper is sufficiently washed. When the filter-paper is to be burned and weighed it must be perfectly freed from the acids by repeated washing.

5. Pour 10 or 15 drops of the filtrate into a test-tube, and add three or four drops of hydrochloric acid. If a precipitate forms it may be of silver; if so, it will grow dark violet on exposure to daylight, or more rapidly and darker in sunlight. To test more quickly, add strong ammonia — 30 to 40 drops; the precipitate dissolves after a short time; if it does not dissolve, then it is lead. Filter and test on charcoal with the blowpipe; if it gives, with inner flame, a bead and yellow incrustation around, it is lead. If none of the above results are given, and yet there is a precipitate, then it is mercury. To prove this, add a solution of carbonate of potash and digest; it turns black; filter and place in a glass tube, heat gently with a blowpipe; it volatilizes and condenses on the sides; examine with strong lens — it is mercury.

6. If hydrochloric acid produces no precipitate, though in excess and heated there is neither lead, silver, nor mercury in the assay, and it is not necessary to treat the ore for either, but proceed to the next step. It will be seen why nitric acid was added to the assay, as in No. 2. Hydrochloric acid would have prevented these tests as given, but we are now prepared for the next metals, with three less to look for, or with a certainty as to the presence of one or more of the three.

7. The whole assay, or its solution, may now be used. If any precipitate formed in the test-tube, treat the whole solution with hydrochloric acid, heat to boiling, and separate the precipitated metal or metals in the whole, as in the test-tube, by filtration. Wash, set the filter paper aside under cover of paper to dry, and pass hydrogen sulphide through the filtrate slowly until it smells plainly of the gas.

8. As this gas is frequently used, make a simple and

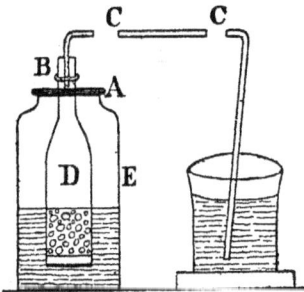

FIG. 34.

cheap apparatus so that there will always be a supply on hand. Cut off the bottom of a long bottle of small diameter, *D,* say about two inches, and fit it into a fruit-jar, *E,* as in Fig. 34.

The top, *A,* should be fitted loosely, so that it may be removed and let air pass through. The cork at B must be air-tight. Fit a small tube into the cork after bending it in a flame; a ¼-inch tube with an ⅛-inch aperture is sufficiently large and is easily bent. If no iron sulphide is on hand, it may be made by heating an inch rod of iron white-hot, and press it into a small piece of sulphur (nut size) and the iron melts readily against the sulphur. Place some cotton in the neck of the bottle, and having fitted a plug of wood with holes in it for the bottom of the bottle, invert it and half fill with iron-sulphide lumps, fasten the wooden plug in the bottom, not very tightly, but in three or four places, so that water can pass easily, and yet the plug be well fixed. Put the bottle in its place, resting in the jar at *A*, and somewhat loosely fastened. This is done after the jar has been half filled with a mixture of equal parts of common hydrochloric acid and rain-water (or well water). Hydrogen sulphide will form immediately, and the gas will pass from this apparatus into the solution of ore in the beaker, so that precipitation will soon take place. If two little blocks of wood are tied against the sides of the india-rubber tubes, *C C*, so as to press the sides together and stop the gas from flowing, the gas being generated pushes the water out of the interior glass *D*, and stops forming, but is ready at any moment to begin as soon as the string around the blocks is removed.

9. After introducing the hydrogen sulphide until the filtrate smells of the gas, filter and wash the precipitate, mark the paper containing it *A*, and put this precipitate aside for the present. This is the precipitate from the hydrogen sulphide.

10. The filtrate. If the strip of platinum shows that it contains some material after evaporation of a few drops, proceed by adding a solution of ammonium chloride (sal-ammoniac), and then liquid ammonia to the filtrate, using about $\frac{1}{15}$ or $\frac{1}{20}$ of the bulk. Then add ammonium sulphide so long as any precipitate is apparent. Let it stand awhile. This precipitate may contain alumina, chromic

oxide, zinc, nickel, manganese, cobalt, and iron as sulphides. It may also contain phosphates, borates, oxalates, and hydro-fluorates of the alkaline earths (barium, strontium, and lime), which will not be considered further.

11. Filter and wash the precipitate. Add a little water to the hydrochloric acid, now to be used in treating this precipitate. Add this diluted acid in sufficient quantity to dissolve the precipitate, and put it aside to digest. If any part refuses to dissolve, it is because there may be present cobalt or nickel, or both; add nitric acid and boil, for these metals dissolve in hot nitro-hydrochloric acid. Filter. Add ammonium chloride to the whole solution, and excess of liquid ammonia. The resultant precipitate may contain alumina, chromic oxide, sesqui-oxide of iron, and the alkaline earths, as phosphates, etc. Dissolve the precipitate in caustic potash solution until all is dissolved that will do so. Filter. The solution may contain alumina and chromic oxide; boil for some time, and if a precipitate is formed, it is chromic oxide; confirm by the blowpipe; it gives a green bead with borax, heightened by fusion with metallic tin or charcoal, which is the blowpipe test for chromium.

12. Next super-saturate the solution with hydrochloric acid and boil with excess * of ammonia; if a precipitate is formed it is alumina. Confirm with blowpipe, as has been shown. What was dissolved by digestion with potassium hydroxide (caustic potash solution) has now been treated. The precipitate may contain iron and more chromic oxide, and the phosphates, etc., of the alkaline earths.

* By 'excess' is meant so much that after stirring with a glass rod, the liquid smells strongly of ammonia.

13. Now proceed with a portion of this precipitate by first dissolving it in as small a quantity of hydrochloric acid as is possible, filter, and add to the solution (made as nearly neutral as possible) two or three drops of ferrocyanide of potash (yellow prussiate of potash in solution) ; a blue precipitate is formed, proving the presence of iron sesqui-oxide. Wash another portion and fuse it in a small crucible with potassium nitrate (pure saltpeter) and sodium carbonate about equal parts. When cold, digest with water; a yellow solution results, which produces a yellow precipitate with acetate of lead, showing the presence of oxide of chromium. This double reaction for chromic oxide (it was found before) is due to the relative quantity of iron present as related to chromium present, which will not be precipitated entirely at one time in the presence of iron, under these circumstances.

14. Next take up the solution filtered off from the precipitate treated in section 11. This solution may contain zinc, manganese, nickel, and cobalt. Digest with ammonium sulphide, wash the resultant precipitate and dissolve it in nitro-hydrochloric acid. It may be dissolved upon the filter by dropping the mixed acids and filtering through into a clean beaker, just as it could have been done in No. 11. This is convenient when the precipitate adheres too tightly to the filter to allow scraping it off entirely. Digest this clear solution with potassium hydroxide (or caustic potash) precisely as in No. 11. This potash may be put into the beaker in small pieces of the stick, in which form it is generally sold.

(a) The solution may contain zinc oxide.

(b) The precipitate may contain manganese, cobalt

and nickel, as oxides. Pass hydrogen sulphide through the solution (a) until the precipitate (white zinc) ceases to fall. Wash and agitate the precipitate (b) with a solution of carbonate of ammonia. The precipitate that now falls is the carbonate of manganese; confirm this by blowpipe. The solution from this last treatment may contain cobalt and nickel oxides. Evaporate it to dryness, re-dissolve in a few drops of hydrochloric acid, and again evaporate to a moist mass and divide into two parts. Heat one portion with borax in the blowpipe flame, a blue bead proves cobalt. Dissolve the other portion in water and slowly add solution of cyanide of potassium; a precipitate is formed which, on continued addition of the cyanide, begins to re-dissolve. On adding hydrochloric acid it is again precipitated. It is nickel; confirm with blowpipe.

15. In No. 9, paper *A* was put aside. This contained the precipitate holding the copper of the ore, if any was present. Digest this with ammonium sulphide (or potassium sulphide). A solution and a precipitate are formed. The precipitate may contain lead, mercury, bismuth, cadmium, besides copper, as sulphides. The solution may contain gold, platinum, antimony, arsenic, and tin as sulphides.

16. Treat the precipitate first, by boiling it with nitric acid. A black or brownish residue remains undissolved. Take a hard, glass tube, and having washed and dried the black residue, put some of it into the tube and heat it. It may act in three ways: (a) it sublimes without change; mercury oxide was present — test with blowpipe; (b) it sublimes, leaving a white powder which, when moistened with ammonium sulphide, turns black, proving it to

be lead sulphate; (c) it sublimes, but as a mixture of mercury sulphide with minute globules of metallic mercury, showing that through some haste or lack of care, mercury as a sub-oxide still remains when it should have been entirely precipitated as chloride of mercury at the first (section 5).

17. Proceed with the filtrate (obtained as stated in No. 16), from the black or brownish residue. Treat this with solution of carbonate of potash and wash the resultant precipitate, and then digest it in cyanide of potassium in excess, while it is moist. This may be done on the filter after changing the beaker, since this filtrate or solution must be kept. The insoluble part may contain lead and bismuth as carbonates; the solution may contain copper and cadmium as double salts with cyanide.

18. Proceed with the insoluble part by boiling it with dilute hydrochloric acid. To one part of the resultant solution add sulphuric acid; the precipitate indicates lead. To·the other part, after concentration by evaporation, add a large quantity of water; a milkiness is produced indicating bismuth.

19. Into the solution (No. 17), after digesting with cyanide, pass hydrogen sulphide — the precipitate, if formed, indicates cadmium — test it with the blowpipe. To the solution add hydrochloric acid, copper sulphide will be precipitated; add a few drops of nitric acid, which will dissolve the copper sulphide, and by adding ammonia slightly in excess the solution has a blue color indicating copper.

20. Next treat the solution mentioned in No. 15. The insoluble part, No. 16, having been separated off as there stated, add acetic acid to the solution, and boil. If a pre-

cipitate be produced, collect a small portion, wash and heat it over a spirit-lamp upon a strip of platinum foil.   If it burns with a bluish flame and leaves no residue whatever, it is sulphur and nothing more may be done as this part of the assay is exhausted.   But if it leaves some residue, then several important elements may be present. Proceed, and to one part add a solution of chloride of tin (proto-chloride with a drop of nitric acid added), a purple color is produced.   To another part add a solution of proto-sulphate of iron — a brown precipitate is produced indicating gold in both cases.

To another part add ammonium chloride solution; a yellow crystalline precipitate falls which indicates platinum.   Arsenic in the ore may be tested by the blowpipe, but if the presence of sulphur, in larger quantity, prevents detection of a small quantity of arsenic, it may be identified thus: Take part of the black or brownish precipitate resulting from the addition of acetic acid, and mix it with three times its bulk of nitrate of potash (saltpeter) and carbonate of soda.   Put this mixture, a little at a time, into a porcelain crucible, in which a mixture of these has been placed and is in fusion over a lamp.   At conclusion, digest the fused mass with pure water; filter; add excess of nitric acid and heat; now add nitrate of silver; filter when cold, and add very dilute ammonia; a brown precipitation on coloring shows arsenic.

Dissolve another portion of the dark precipitate or residue from acetic acid in hydrochloric.   Place a strip of metallic zinc in the solution; a pulverulent deposit takes place on the zinc, indicating antimony.   If more proof be wanted remove the powder to a beaker and digest in nitric acid, when a white precipitate is formed. Digest it with a strong solution of tartaric acid; only a part may be dissolved, but filter; into the clear solution pass hydrogen sulphide, and an orange-colored precipitate is formed, proving antimony.

In the last paragraph it was found that part of the precipitate was not dissolved in the tartaric acid; dry it; place on charcoal with a little cyanide of potassium and carbonate of soda, and turn the inner flame of the blowpipe upon it; it is reduced to metallic tin.

In the above analysis provision has been made for the detection of 16 elements. Of course, if no precipitate or signs appear at any one stage of the analysis, proceed immediately to the next, as it is not probable that any mineral will ever contain half the elements mentioned but the full number is given so as to serve any possibilities.

II. DRY ASSAY OF ORES

As much of the dry assay as may generally be needed is given herewith:

Apparatus required includes crucibles, scorifiers, and cupels. The first two are made of good fireclay. Crucibles holding.a 6 or 8-oz. charge are the best size for field work. They are used for large quantities of ore. Scorifiers are flat, but thick saucers, intended to prepare the ore for the final treatment in cupels, but are only used for reducing rich ore, or reducing the size of lead buttons formed in the crucibles. The cupel is a little saucer of bone-ash, intended to be used on the floor or bottom of a heated muffle of the assay furnace. The muffle is a clay oven of small dimensions, intended to protect the scorifier or cupel from the fuel of the furnace.

An assay furnace may be made of sheet-iron; it should be 15 inches in diameter, with a grate near the bottom, and lined with either ordinary or fire-brick. In the accompanying illustration, Fig. 35, is given the general form of one that has been used for years. A plain sheet-iron cylinder 18 inches high and 15 inches diameter, with draft-hole at A, muffle-hole at B, and pipe-hole at C,

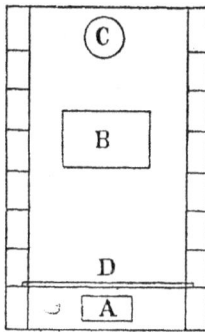

Fig. 35.

and lined with brick, will answer all purposes. The hole at C must have a collar and pipe either for a chimney or it must enter a stack. B must be provided with a flanged door, as also the draft hole A. The top may have, loosely laid on, only a square sheet of heavy iron, and the whole placed upon a flat stone or few bricks. Several bars (¾ in. square) of iron fitted into the bricks will answer where there is no foundry at hand to cast a grating D. Charcoal or coke may be used, or, where the draft is strong, a hard coal.

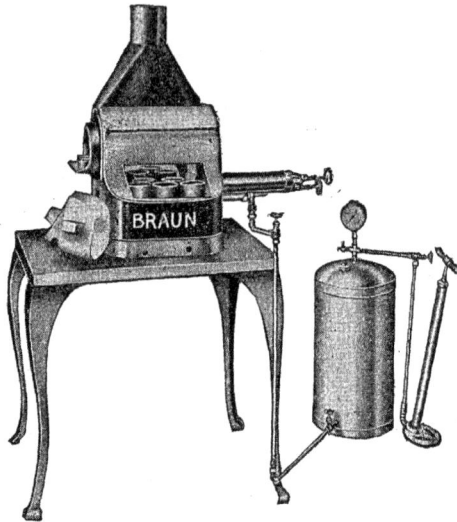

FIG. 36.

The above illustration shows a sectional view of a rotary-flame combination furnace with Cary hydro-carbon burner, as supplied by Braun-Knecht-Heimann Co. of San Francisco. This furnace is constructed in such a manner that the muffle is placed above the crucible compartment, and in this respect differs from this firm's other types of combination muffle and crucible furnaces. In this furnace a large crucible compartment is obtained, and the different portions of this compartment are, to a limited extent, of different temperatures. The burner hole opening of the furnace is in the center of one side, and the flame from the burner is forced toward the opposite wall of the furnace, where it is deflected to both sides of the chamber by a brick of special shape, which forces the flame in a rotary motion around both sides of the crucible compartment. This brick is always in the hottest part of the flame, thus saving the fire-clay lining of the furnace, and, being separate, can easily be replaced at a slight cost.

The draft for cupellation is obtained through an opening 1¼ inches diameter made in the jacket and lining at a point opposite the hole in the rear end of the muffle, and the amount of air admitted to the muffle is controlled by a slide fitted to the jacket. An opening larger than the inlet is provided just inside of the jacket and directly over the open end of the muffle. This outlet is opened and closed by withdrawing or inserting the fire-clay muffle-plug. The heat rising from the crucible chamber creates by suction a draft through the full length of the muffle when the inlet and outlet apertures are open, thus securing rapid cupellation. The jacket of this furnace is made of sheet iron. The complete outfit for 6-F or 4-G crucibles, weighs 205 pounds, or packed 310 pounds, and costs $90.

A handy portable furnace, weighing only 30 pounds, is the Hoskins' combination No. 5, made by E. H. Sargent & Co., of Chicago. This takes a crucible as large as 3 by 5 inches or two smaller ones, and a muffle 6 by

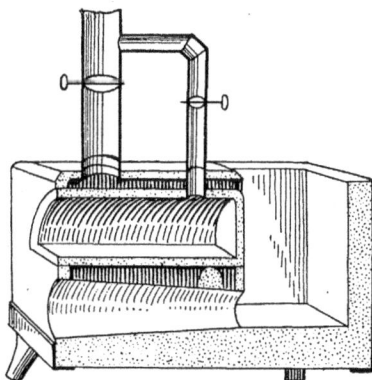

FIG. 37.

3 by 2½ inches, and is very compact. Heat is applied by a Hoskins' blowpipe No. 3, with a 1-gallon capacity container, using 74-degree (or thereabouts) gasoline.

If it is desired to ascertain the amount of an ordinary metal, such as iron or lead, pulverize the ore to pass a 40-mesh screen, weigh it, mix with charcoal on paper, and put the mixture in the bottom of the crucible, cover with charcoal an inch or two deep, drop in two or three pieces of brick, and place the crucible in the hottest part of the fire. Cover all with coal, and gradually increase the heat and keep it nearly at white heat for half an hour. Then draw it out, and tap the crucible down on a stone to settle the melted button. When cool, take out the contents, and the metallic iron or lead will be found with its slag attached. Clean the button and weigh it, then

$$\frac{\text{weight of resulting button}}{\text{weight of sample of ore}} \times 100 = \text{percentage of metal}$$

in the ore.

Any scales that weigh from ¼ to ½ lb. or more will serve for rough work in the field. The cheapest and lightest scale is one used for weighing letters, which weighs from ¼ to 12 oz.; but a better scale is a light spring balance, weighing up to 2 lb., and divided into ½ and ¼ oz.

The sample may be weighed best by laying it on a sheet of paper, turning up the edges, and tying them with a piece of string, which can be hooked on to the scales.

For more delicate work, a small pair of scales weighing to ¹⁄₁₀₀th of a grain is sufficient. Such scales may be bought at any chemical warehouse, made to pack and carry with ease and security. When in a regular laboratory, the scales weighing to 0.0077 grain, or half a milligram, will save chemicals, time, and work; but unless the analyst has a true average sample of the ton of ore, the smaller the amount of ore used the more likely is the assay to prove deceptive when proportioned to the ton.

In weighing the ore it is well to make use of the conventional assay-ton weights, as by this system the number of ounces of precious metal in a ton of ore may be known, according to the number of milligrams, etc., the button of precious metal weighs. The assay-ton (A. T.) weighs 29.166 grams (450 grains), or 29,166 milligrams. If 1 A. T. of ore yields a button of 1 milligram, a ton of ore yields 1 oz. troy of precious metal.

SAMPLING AND PULVERIZING

Sampling is of the utmost importance, and is discussed fully on page 41, especially with gold ore, as a very small quantity of gold, more or less, makes a great difference in the estimated value of a vein or deposit. Do not take selected small pieces, but an average sample of the mineral deposit. Pulverize the specimen carefully in a mortar, or, in the absence of the latter, break the ore up into a few pieces, wrap the latter in cloth or paper, and powder between two hard rocks. To prevent fragments of ore from flying out of the mortar, cover the latter with a piece of paper with a hole in the center for the pestle to pass through. Quartz and similar substances will be rendered easier to crush by first being heated and then thrown into water, but this should not be done if sulphides be present. Crushing to 60-mesh is fine enough. Smaller particles may be lost or separated in the crucible. If a sieve is not available, obtain a piece of silk bolting-cloth.

Cleanliness is an essential, otherwise samples may ' salt ' one another, so when fragments or ore adhere to the mortar, a little pulverized charcoal should be stirred about in the mortar.

Gold and Silver Ores. If rich, these require preparation in the scorifier. Powder the ore. Take 50 grains, 500 to 1000 grains of granulated lead, according to the probable amount of silver, and 50 grains of borax. Mix the ore with half the lead and place the mixture in the scorifier, spread the other half of lead over the contents, and finally spread the borax over all. Put the scorifier in the muffle, close the door, and heat up to fusion; then the door should be partly opened, the heat increased, until the oxidized lead (litharge) covers the scorifier. Take the latter from the muffle and pour the contents into an iron cavity or mold, separate the button and hammer it into the shape of a cube. It is now ready for cupellation, as it contains all the gold and silver.

By cupellation the lead is simply separated from the gold and silver, this being effected both by absorption in the bone-ash and oxidization. Cupels may be made, but they can be bought so cheaply that it is seldom worth the trouble to make them.

Push a cupel into the heated muffle, place the cube of lead in the cupel with tongs, and heat up until the lead melts. Watch the lead gradually reducing until it is the size of the silver it contains, when the surface will become suddenly bright, and nothing remains but the silver containing the gold. Withdraw the cupel and cool and weigh the button or prill. The gold and silver must be separated by the wet proces, thus: Dissolve the button in dilute nitric and heat until the acid boils. A dark powder precipitates. Filter off the dark powder, which is gold, and precipitate the silver by solution of common

table salt or by hydrochloric acid. After all is thrown
down, drop into the white precipitate some pieces of zinc,
add more hydrochloric acid — hydrogen gas is generated,
which reduces the white silver chloride to powdered metallic silver. The gold and the silver may now be melted
in separate crucibles, weighed and compared with the
amount of ore used. This is rather a cumbersome method
of finishing an assay, a much easier way being to weigh
the button after cupellation, dissolve out the silver with
nitric acid, wash and dry the gold remaining, and weigh
it, the difference between the two weighings being the
silver.

In these trials the lead should first be cupelled for its
silver, and that subtracted from the silver found, as nearly
all lead contains some silver.

If it should be more convenient to melt the ore in a
crucible rather than in a scorifier, use the following flux:
If the ore is composed chiefly of quartz, pulverize, take
100 to 500 grains of ore, 500 grains of red lead, 20 to 25
grains of charcoal powder, 500 grains of carbonate of
soda and borax together — the more silica the more carbonate of soda, the more metallic bases the more borax.
Place a little borax over all and melt until all is liquid,
requiring about 20 minutes; withdraw, extract the lead
button when cool, hammer up to a cube and cupel. Separate the gold and silver as before, but the amount of
silver must be three times that of the gold, and if there
is reason to believe that there is not this amount, some
silver must be melted with the button, since the separation
will not otherwise be complete.

A regular process of assaying gold quartz is to take
200 grains of ore, 500 of litharge, 6 of charcoal or 20 of

flour, and 500 of carbonate of soda; or, 200 grains of ore, 200 of red lead, 150 of carbonate of soda, 8 of charcoal, and 6 of borax. Mix and put into a warm crucible, and cover with half an inch of common salt. Fuse in a hot fire for 30 minutes; either pour into a mold or cool and break the pot; and clean the button with a hammer.

If a quartz is pyritic, take 100 grains and calcine dead without sintering, add 500 grains of red lead, 35 of charcoal, 400 of borax, and 400 of carbonate of soda, cover with salt, and proceed as above. In each case cupel the button.

As the bone-ash of which the cupel is made can absorb its own weight of metallic oxides, the cupel chosen should always exceed the weight of the button to be heated, so as to have an excess. Boil the gold prill obtained from cupelling in nitric acid, which dissolves the silver and leaves the gold pure.

The above formulas are open to modification by the operator according to the apparent richness or poverty of the ore to be treated, and the presence and character of the basic impurities. In case there are oxides, a reducing agent is required; and if sulphides, an oxidizing agent. As a rule, employ a weight of litharge twice that of the ore, and of carbonate of soda the same as the ore. These reagents are added to control the size of the lead button, and to obtain one of suitable size for cupelling.

The presence of metals in the ore is indicated by cupel stains as follows: Antimony — pale yellow to brownish-red scoria; sometimes the cupel cracks. Arsenic — white or pale-yellow scoria. Cobalt — dark-green scoria and greenish stain. Copper — green or gray, dark red or brown. Iron — dark red-brown. Lead — straw or orange color. Manganese — pink and dark bluish-black stain. Nickel —.greenish stain; scoria dark green. Palladium and platinum — greenish stain; the button will be very crystalline. Tin — gray scoria; tin produces ' freezing.' Zinc — yellow on cupel; the cupel is corroded.

Lead Ore. To ascertain the amount of lead in galena, the common ore of lead, charge the crucible with the powdered ore, carbonate of soda — two or three times the weight of the ore — three or four long nails on top to absorb the sulphur, and a covering of salt or borax. Heat to redness about 20 minutes. Pour the contents into a mold, and separate the button from the slag.

$$\frac{\text{Weight of button}}{\text{Weight of ore sample}} \times 100 = \text{percentage of metal}$$

As galena always contains more or less silver, the resulting button should be assayed for the precious metal in the cupel. As the latter does not absorb much more than its own weight of lead, the button may have to be divided into two or more portions, and each of these cupelled separately, or else scorified to a size suitable for cupellation.

Copper Ore. The wet assay is better than the dry, especially that by the burette, which will be given later on under copper.

Tin Ore. If it is mixed with iron or copper pyrite it should be powdered and roasted, and then mixed with one-quarter of its weight of charcoal and subjected to great heat in a crucible for about 20 minutes. Tap it as in an iron assay, let it cool, and pick out the button or buttons, or pour it out while melted.

It may also be reduced by melting the powdered ore with cyanide of potassium, 100 grains of ore to 600 grains of cyanide. Cool, and extract button. This ore is hard, and should be powdered to 60-mesh.

Mercury. These ores are easily reduced by simply heating and condensing the vapors in a cold bath such as in using a retort and cool receiver.

Antimony. Place about 2000 grains of powdered ore in a crucible having a hole bored in the bottom, and the hole stopped loosely with a piece of charcoal. Put this crucible into another half-way down. Then lute on the lid with clay and put clay around the junction of the two, and put live coals around the upper crucible by placing some broken bricks around the lower one on the grate, to keep the coals away from it. The antimony will melt and leave its gangue-rock in the upper crucible while the lower one will receive the melted metal.

Bismuth, zinc, manganese, nickel, cobalt, and other metals should be reduced or analyzed by the wet process already given in this chapter.

In the following chapters will be discussed the metals and minerals of commercial importance.

What to take on a prospecting trip is a matter of individual judgment, and of course depends on the technical ability of the person or party going out, financial resources, means of transportation, and length of the season. An old-time prospector's needs were few, but the scientifically arranged outfit is fairly elaborate and costs a good deal. In 1903, the reviser of this edition went with an exploration party into western Borneo, Dutch East Indies, taking a complete portable testing plant, as well as the usual panning outfit carried by prospectors. This cost then about $250. The following list, suggested by The Denver Fire Clay Co. of Colorado, is somewhat similar to that taken on the Borneo trip. It costs $160, and weighs from 300 to 350 pounds, packed:

1 Portable button balance and weights.	3 Beakers and covers.
1 Pulp balance and weights.	1 Blowpipe, Plattner's.
1 Furnace (Burro or Brown) or,	3 Funnels.
	1 Package filter-paper.
	1 Button brush.
1 Case gasoline furnace with blowpipe tank.	1 Wash-bottle.
2 Muffles.	6 Parting-flasks.
200 Scorifiers.	1 Tripod.
50 Crucibles.	6 Annealing-cups.
1 Quart mortar and pestle (iron).	2 Hammers.
	4 Lb. litharge.
2 Pairs tongs.	5 Lb. soda bicarbonate.
1 Magnifying lens.	1 Lb. argol.
1 Lead mold.	1 Lb. muriatic acid, com. pure.
1 Cupel mold.	1 Lb. nitric acid, c. p.
1 Magnet.	10 Lb. bone-ash.
3 Pairs pliers.	2 Lb. borax glass.
1 Spatula,	¼ oz. silver foil, c. p.
Glass rod and tubes.	½ Lb. rolled lead, c. p.
1 Glass alcohol lamp.	10 Lb. granulated lead, c. p.
1 Sieve, 60-mesh.	1 Pt. alcohol.
	2 Lb. lead flux.

The Braun-Knecht-Heimann Co. of San Francisco suggest the following:

The Pritchard chemical process enables any one, without the knowledge of chemistry, to make accurate assays of gold, silver, copper, and lead. It is especially recommended for prospectors and small mine-owners, being simple and easily operated, and accurate. The outfit is entirely portable, weighing but 20 lb., and costs $60. The prospector or small mine-owner must not only be able to know gold and other minerals when he can see them, but he must be able to detect them in base and low-grade ores, and determine their commercial value. The prospector who can do this in the field can often save the cost of the entire outfit by being able to make assays on the spot. The small mine-owner can make many assays at the ordinary cost of one. He soon learns just what character of ores are of value, and is in-dependent of all mistakes. He knows when and where to work, and when to move on. The prospector can earn his own grub-stake as he goes along by making assays.

FIG. 38.— BALANCE AND SCALES.

Chemicals sufficient for 150 assays are furnished with each outfit.

The nitric acid and sulphuric acid can only be shipped with the outfit when sent by express, as they are not acceptable by parcel post.

Gold Assay — The operation is extremely simple. The gold is extracted by a solution, and precipitated from the solution by specially prepared mercury. The mercury is dissolved out, leaving a pure, porous gold button or bead. The value of the gold button is now determined by weighing it on the button balance. This is simply constructed and accurate, and is sensitive to one one-hundredth of a milligram. The balance is enclosed in a mahogany case with transparent celluloid face to keep out any moving air when weighing.

The pulp scale has a bearing which sets in the end of the case as shown in the cut — Fig. 38.

Silver Assay — The silver and all of its salts are dissolved from the ore by combining the gold solvent with the silver sol-

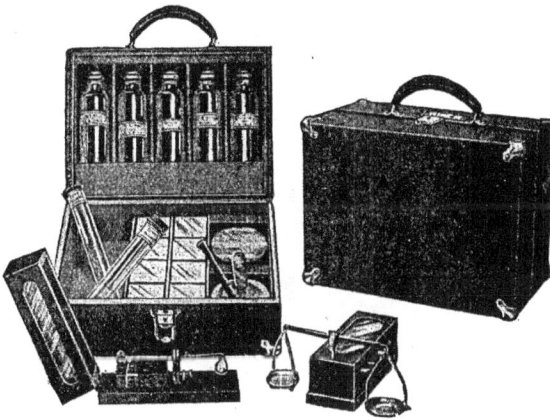

FIG. 39.— PORTABLE TESTING OUTFIT.

vent. The silver is precipitated on zinc, re-dissolved, and its value determined by what is known as the volumetric process; that is, the amount of silver is determined by the amount of silver precipitant that is required to precipitate it. The reaction is plain and distinct. Distilled water is required for making the volumetric assay.

Copper Assay — The copper is dissolved from the ore with acid. The percentage of copper is determined accurately by examining its blue ammoniacal solution through a specially designed copper gage. It requires only 30 minutes to make a complete assay.

Lead Assay — The lead in the ore is reduced to a lead sponge
• by acid and specially prepared zinc. This sponge is re-dissolved
and precipitated with a lead precipitant, and again reduced to a
pure lead sponge, which is pressed, dried, and weighed. Approximate assays can be made in 30 minutes, and an accurate assay
will require about one hour.

Book of Instrucions — A book of instructions is furnished
with each set of apparatus, accurately and plainly explaining

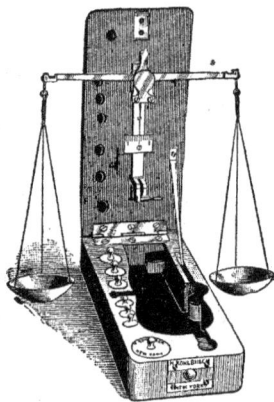

FIG. 40.

every detail of the processes. With this book of instructions
one needs no assistance from anybody.

The Pocket style balance button is a useful and accurate instrument. Its sensibility is 1/10 (0.1) milligram, and it has a
five-inch beam. When closed the case measures 6 by 2¾ by 1¼
inches. Weights are 10 grams down to 1 milligram, neatly fitted
in box as shown in cut; shows 5 divisions for 1 milligram and
½ division for 1/10 milligram. The cost is $20.

A complete portable assay outfit, as supplied by Braun-Knecht-
Heimann Co., San Francisco, is as follows:

1 Braun rotary-flame combination furnace No. 43.
1 Cary burner.
1 Tank outfit, capacity 2 gallons.
1 Set portable button and pulp balances.
1 Set weights, 1 gram to 1/10 mg.
1 Set assay-ton weights.
1 Pair pincets for weights.
2 Riders, 1 milligram.
1 Camel hair brush.
2 Camel hair pencils.
1 Pipe wrench, 8 inch.
2 Extra muffles for furnace.
1 Pair cupel tongs, 30 inch.
1 Pair scorifier tongs, 30 inch.
1 Pair crucible tongs, 30 inch.
1 Pouring mold, 3-hole.
1 Pair forceps, 5 inch.
1 Slag hammer 16 ounce.
1 Iron cupel mold, 1¼ inch.
1 Wood mallet.
6 Roasting dishes, 3 inch.
25 Scorifiers, 2 inch.
100 Crucibles, 20 gram.
1 Prospector's mortar and pestle, large size.
1 Gold washing pan, 16 inch.
1 Metal frame sieve, 8 inch diameter, 60-mesh.
1 Metal frame sieve, 8 inch diameter, 80-mesh.
1 Mouth blowpipe, 10 inch.
1 Alcohol lamp, 4 ounce.
1 Bead anvil.
1 Button hammer, 16 ounce.

1 Button brush.
1 Pair curved pliers.
1 Kerosene oil stove.
6 Porcelain annealing cups, No. 0.
1 Pair tongs for annealing cups.
6 Parting flasks, 20 cc.
1 Wash-bottle complete, 8 ounce.
1 Wash-bottle complete, 16 ounce.
½ Pound glass tubing, ¼ inch.
1 Three-cornered file, 4 inch.
3 Feet rubber tubing, ¼ inch.
1 Dozen test-tubes, assorted sizes.
1 Test-tube brush.
1 Test-tube clamp.
1 Horseshoe magnet, 4 inch.
1 Pocket magnifying glass, 1¼ inch.
1 Yard sampling cloth, 36 inch.
12 Sheets glazed paper.
1 Spatula, 3 inch for weighing.
1 Spatula, 6 inch, for mixing.
50 Paper ore sample-bags, metal fastener, 4 ounce.
1 Pound hydrochloric acid, c. p.
1 Pound nitric acid, c. p.
1 Pound argol.
5 Pounds borax-glass powdered.
½ Pound charcoal powdered.
10 Pounds litharge, free of silver.
2 Pounds granulated lead, c. p.

1 Pound sheet lead, c. p.
1 Pound quicksilver.
1 Pound potassium nitrate, coml.
½ Ounce silver foil, c. p.

15 Pounds sodium bicarbonate.
5 Pounds bone-ash.
1 Quart alcohol denatured for burning.

In addition to above both kerosene and gasoline will be necessary. These have been purposely omitted, as it is presumed that they will be obtained at destination. This complete outfit, packed for shipment, costs approximately $300.

CHAPTER VI

GOLD, SILVER, AND PLATINUM

These are known as the precious metals, and are worth at the time of writing $20.67, $1.17, and $130 per ounce, respectively. Gold is the basis of credit and exchange, and is fixed at this value; but the other two metals fluctuate from day to day, depending on the demand for coinage for silver and for the arts for both it and platinum. While about 90% of the world's gold is obtained from gold ores, only 30% of the silver is recovered from silver ores, the remainder being a by-product from copper, lead, and zinc ores, also some from gold ores. Platinum is nearly always associated with gold in alluvial deposits, and in Colombia the proportions approximate 75% platinum and 25% gold. It will be seen that the search for base-metal ores is worth-while for their silver contents also.

GOLD

The yellow metal is widely distributed throughout the world, but its deposition in commercial quantities is limited. Ore that carries as little as 1 pennyweight per ton may be worked on a large scale at a profit, but generally the quantity should be several times that amount. The factors that determine what is pay-ore depend upon local economic conditions.

The great gold-bearing regions of the earth are the Transvaal in South Africa, the western United States, Australia, Siberia, Ontario, and Mexico. In each of these areas are small districts in which gold is mined. In addition to the countries named, gold is found in Europe, India, Central and South America, but only in a few places is it concentrated to any extent, yet all of these countries are worth further prospecting.

Gold exists almost exclusively in the metallic state, either in place in quartz with pyrite, arseno-pyrite, galena,

 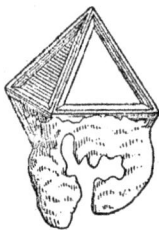 

FIG. 41.   FIG. 42.   FIG. 43.

and many other minerals; or as grains and nuggets in the sands of rivers or in alluvial soils.

Gold crystallizes in the isometric system. The occurrence of well-defined crystals is, however, rare, but they have been found in lodes in California, Australia, and Brazil. The usual forms are cubes and octahedra, with rod and plate-like forms. The crystal faces are frequently striated, and the crystals are generally rounded. Figs. 41 and 42 represent gold crystals. Twinning gives rise to dendritic and reticulated groups, but individual crystal faces and edges are usually very small. Moss-

gold, leaf-gold and wire-gold are other forms. Fig.
43 shows the finest gold dust magnified 700 times,
and Fig. 44 a reduced illustration of a lump of gold
that was found at Forest creek, Victoria, Australia. It
weighed more than 30 pounds, and was 11.33 inches long
and 5.15 inches wide. The largest nugget of gold ever
found was at Ballarat, Australia. It was just under the
surface, and weighed over 191 pounds, being 20 inches
long and 9 inches wide. This was called the 'Welcome
Nugget.'

The specific gravity of gold is 15.6 to 19.3, according

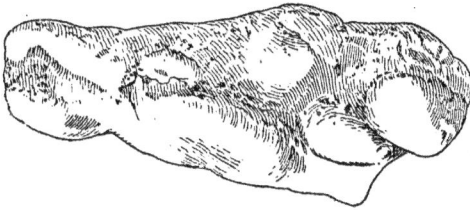

FIG. 44.

to the amount of alloy; hardness 2.5 to 3.0. It is the
only yellow, malleable element found in the natural state.
Its color varies from pale to deep yellow. In some localities it often has a very light color, but it presents the
same color from whatever direction it is looked at, and
to the prospector this is a guiding test. One of the
most important and useful accomplishments for gold
exploitation is 'an eye for color.' Native gold possesses
a peculiar color which is readily recognized, although it
may be alloyed with silver or copper, and its color will
instantly distinguish it from iron or copper pyrite.

Gold grains will always flatten when struck with a

hammer or between two stones, whereas other minerals similar in color will break into fragments. If the doubtful particle is coarse enough, take a needle and stick the point into the specimen in doubt; if gold, the steel point will readily prick it; if pyrite or yellow mica, the point will glance off or only scratch it.

Under the blowpipe, on charcoal, gold may melt, but on cooling it always retains its color. Any other mineral will lose color, become blackened, or will be attracted to the end of a magnetized knife-blade, showing that the unknown substance contains iron.

Gold imparts no color to boiling nitric acid. It will not dissolve in nitric or hydrochloric acid separately, but it does dissolve in the two when combined, which is called aqua regia; proportions: one of nitric to three of muriatic (hydrochloric).

Gold is only mechanically mixed with pyrite, and if nitric acid be poured on a sample it will be dissolved, leaving the gold free. To detect gold in pyrite put a few drops of mercury into a porcelain crucible, place a perforated piece of cardboard in the crucible so that it rests a short distance above the mercury, place a small package of pyrite over the hole in the cardboard, heat the crucible for some time and watch with a pocket-lens for the appearance of white stains of gold amalgam, which on rubbing with a brush or feather becomes lustrous.

Alloys of gold. There are several of these, but their commercial importance is slight. Gold amalgam is found in California and Colombia. It contains 40% gold and 60% mercury.

Bismuth-gold contains up to 64% gold and 34% bismuth. Electrum is a natural gold-silver alloy, with specific gravity between 12.5 and 15.5. The color varies from yellowish-white to pale yellow, depending on the quantity of silver present.

Compounds of gold. These are the tellurides.

Tellurium is a rare element occasionally found native, for instance, in California and Colorado, but more commonly in combination with gold, silver, lead, and bismuth. Native tellurium is tin-white, crystalline in structure, brittle, and therefore easily reduced to powder. Its specific gravity is 6.1, and hardness, 2.5. It is very fusible, volatilizing almost entirely and tinging the blowpipe flame green. It leaves a white coating on charcoal, and is soluble in nitric acid. The metal is of little commercial value. It is rarely saved at works where ore is roasted, but is recovered at refineries where base metals are treated electrolytically.

The tellurides comprise a small, but interesting group. The most important deposits are at Cripple Creek, Colorado, and Kalgoorlie, Western Australia, where the sulpho-telluride ores have yielded nearly $400,000,000 of gold at each center. Tellurides have been extracted in large quantities at Goldfield, Nevada; and in lesser amounts in California, Transylvania, and Ontario.

The presence of a telluride is recognized by the purplish-red color of the solution on heating the powdered mineral in a tube, closed at one end, together with charcoal and carbonate of soda, and adding hot water. When boiled in sulphuric acid a telluride yields a pinkish solution. The most important tellurides are:

Calaverite. First found in California, but is in large quantities at Cripple Creek and Kalgoorlie. It occurs massive; has a pale bronze-yellow color, and a yellowish-gray streak; hardness is 2 to 3; specific gravity is 9.34; and it contains 56% gold and 44% tellurium.

Nagyagite, foliated or black tellurium. This is a sulpho-telluride of lead and gold with antimony, containing up to 60% lead, 9% gold, and 30% tellurium. Streak — blackish lead-gray; color — blackish lead-gray; luster — metallic; specific gravity — 7; and hardness — 1 to 1.5. It is sectile, flexible in thin laminæ, and exists in granular or foliated masses. If the mineral is treated for some time in the oxidizing flame, a malleable globule of gold remains. This cupelled with a little assay lead assumes a pure yellow color. Nagyagite forms a valuable gold ore in Nagyag, Transylvania.

Hessite. This is a telluride of silver, often high in gold, carrying up to 59% silver and 37% tellurium. Its streak is iron-black; color, lead to steel-gray; luster, metallic; and sectile, brittle. It forms cubic masses of fine-grained texture. Specific gravity, 8.4, and hardness, 2.5 to 3. Before the blowpipe it fuses on charcoal to a black globule; this heated in reducing flame presents on cooling white dendritic prints of silver on its surface; with soda is reduced to a globule of silver.

Petzite. A telluride of silver and gold. Streak is iron-black; color, steel-gray, iron-black, sometimes peacock tarnish; luster, metallic. Sectile, brittle. Specific gravity, 8.7 to 9. Hardness, 2.5 to 3. Forms cubic masses of fine-grained texture, like hessite, which it resembles in most physical properties, but is much denser. In one locality in Colorado it forms one of the principal minerals in a group of quartz veins in porphyries traversing coarse granites, and occurs in rounded masses, sometimes implanted on iron pyrite and irregular crystalline aggregates, which are occasionally coated with encrusting pseudo-morphs of gold. The Red Cloud mine in Colorado yields a mineral consisting of 33.49% tellurium, 40.73% silver, and 24.60% gold.

Sylvanite or **graphic tellurium.** This is a telluride of gold and silver, averaging 60.82% tellurium, 27.32% gold, and 11% silver at Cripple Creek and Kalgoorlie. Its streak is steel-gray to silver-white; and color, steel-gray to silver-white, and sometimes nearly brass-yellow. Luster — metallic; sectile — brittle in thin laminæ; specific gravity — 8; hardness, 1.5 to 2. Colors the flame blue or bluish-green, giving a white incrustation and a dark-gray bead, which can be reduced alone after long blowing, or more quickly with soda, to a yellow, malleable, metallic bead of silvery-gold. In California, sylvanite occurs in narrow veins traversing porphyry. It is called graphic because of the resemblance in the arrangement of the crystals to writing characters.

Other interesting tellurides are müllerine, a brass-yellow telluride of gold, silver, antimony, and lead; kalgoorlite and coloradoite, a telluride of mercury; and rickardite, a telluride of copper.

It is not always a trustworthy indication that particles are gold because they will not dissolve in nitric acid; some that are apparently gold-colored will not dissolve in it, and yet contain not a trace of gold.

The simplest instrument for the identification of gold and estimation of the value of auriferous material in which the gold is in a free state, is the ordinary miner's pan. This is a circular dish of black iron, about 12 inches wide and 3 inches deep, with sloping sides, pressed out of one sheet. There should be a slight indentation all round where the sides join the bottom, so as to afford lodgment for the gold, also a riffle near the top at one side. A frying pan free from grease will serve well if nothing else is available. The South American *batea*, made of hard wood in a solid piece, and hollowed out like a shallow funnel, is a good implement when in capable hands; so is the *dulong*, used by the natives in Borneo. Another good substitute for the pan is a kind of magnified shovel without handle made of linden wood, and provided with a vertical wall on three sides. The wooden implements should be slightly charred on the surface to

FIG. 45. A Pan.

show up the gold grains, and should not have been used to hold mercury or amalgam. In washing tin ore, the Cornish used an ordinary shovel with great skill.

The object of 'panning off,' as the operation with the pan or batea is called, is to settle and collect at the bottom of the pan the heaviest particles in the material being tested. Simple as the process of panning appears to be, dexterity is only acquired by considerable practice. The operation is as follows:

The pan is about two-thirds filled with the gravel to be washed, and is then immersed in water, either in a hole or in a rivulet, of such a depth that the operator can conveniently reach the pan with his hands while it rests on the bottom. The object of this is to give him free use of both his hands for stirring up the mass, so that every particle may become thoroughly sodden and dis-integrated. Of course the pan may be held in one hand, and its contents stirred with the other, but the disad-vantages of such a method are obvious.

When the dirt has become thoroughly soaked, the pan is taken in both hands, one on either side, and a little inside of its greatest diameter, and without allowing it to emerge from the water, it is suspended in the hands, not quite level, but tipping somewhat away from the person. In this position it is shaken so as to allow the water to disengage all the light earthy particles and carry them away. The movement may be described as ellip-tical. When this has been concluded there will remain in the pan varying proportions of gold dust, heavy sand, lumps of clay, and gravel stones. These last ac-cumulate on the surface, and are picked off by hand and thrown aside. The lumps of clay must be crumbled and reduced by rubbing, so as to be carried off by the water during the next immersion of the pan. A peculiar turn of the wrist is required to allow the muddy waters to escape in driblets over the depressed edge of the pan, without exercising so much force as to send the lighter portion of the gold after them. At last nothing remains in the pan but the gold dust, with usually some heavy black sand and a little earthy matter. By the final care-ful washing with plenty of clean water, the earthy mat-ter can be removed completely, but the heavy iron-sand cannot be got rid of by any method based upon its specific gravity relative to that of gold.

To remove the sand, one of two simple methods has to be adopted. If it be magnetic, as is usually the case, it may be eliminated to the last grain by stirring the mass carefully with a powerful magnet, care being taken that no particles of gold become mechanically suspended among the black sand.

Where this is ineffectual, recourse must be had to blowing. For this purpose the gold colors and iron-sand is dried, and small quantities of it are placed in an apparatus called a 'blower' — a sort of a shallow scoop, made of tin and open at one end. Holding the blower with its mouth pointed away from him, and gently shaking it so as constantly to change the position of the particles, the operator blows gently along the surface of the contents, regulating the force and direction of his breath so as to remove the sand without disturbing the gold. Where water is available, a pan is the most efficient implement.

Panning such a large quantity as two-thirds of a dish is all right to indicate the presence of gold in rock or gravel, but if one desires to get an approximate idea of how much is in a ton of free-milling quartz, it is best to crush a pound to say 20-mesh, and pan that. Then, as 1 pound is a 2000th part of a ton, experience will tell the prospector how much to expect in a ton of such ore, if it has been fairly sampled. Men have been known to check assays very close by this method on such ore. They are soon able to judge how many grains are in the 'tail' of gold left in a pan, and multiplying by 4 cents for each grain they know the value of a pan, which then must be multiplied by 2000 to equal the value of a ton.

A crude apparatus formerly much used in California and Australia is the cradle or rocker. This, as shown in Fig. 46, is a trough 7 feet in length and 2 feet wide. Several bars are nailed across the bottom at equal distances, and at the upper end a sieve is fixed a foot above the bot-

tom. The whole is mounted upon rollers. Four men
are required to operate the apparatus; one digs the
earth, the second shovels it on to the sieve, the third

FIG. 46.

pours water on the earth in the sieve, while the fourth
keeps the machine continually moving on the rollers.

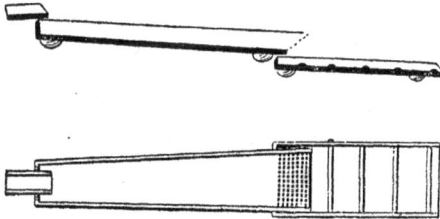

FIG. 47.

One man can work the rocker alone, but the operation
is necessarily intermittent. The large stones washed out
are removed from the sieve by hand, at the same time the

water washes the smaller material through, which is slowly carried towards the lower end of the trough by its slight inclination. Thus the flow of the water tends to keep the earthy particles in suspension so as to allow of their being washed off, while the heavier portions of gold are obstructed in their flow, and retained by the cross-bars fixed to the cradle-bottom, and later are collected.

A more efficient apparatus is the long tom, Fig. 47. A tom that will serve the purpose of the prospector can easily be made at the place where it is decided to test the ore of a newly-discovered deposit. An outfit should contain carpenter's tools, including one or two adzes for giving a smooth plank surface to the side of the timber that forms the floor or sides of the tom. A rough but efficient machine can be constructed in a short time.

The tom consists essentially of two separate troughs, as shown in the figure. These are placed on an incline, or given an inclination by log or rock supports. The Californian tom is 12 feet long, 20 inches wide at the upper end, widening gradually to 30 inches at the mouth. A stream of water flows in by the spout just over the place where the dirt is introduced into the upper box or tom proper. The dirt is thrown in continuously by one man, while a second stirs it about with a square-mouth shovel, or a fork with several blunt prongs, which is useful for throwing out heavy boulders and for tossing back undissolved lumps of clay against the current. The lower end of the tom is cut off obliquely, so that the mouth may be stopped by a sheet of perforated iron. This should be punched closely with half-inch holes — or smaller if the pay dirt is fine — about 20 inches square.

The apparatus being placed on an incline of generally 12 inches, the materials all gravitate with the water towards this sloping grating at the mouth, everything passing through it except the large stones, which gather on the grating, and are removed as often as necessary. Beneath this grating stands the riffle-box, into which falls all the fine sand and gravel, including the gold. The riffle-box, like the tom proper, is made of rough plank, and is also placed on an incline, but only just so that the water passing over it will allow the bottom to become and remain covered with a thin coating of fine mud. In this way the gold and a few of the heaviest minerals will find their way to the bottom and rest there, especially by the help of the riffle-bars. Sometimes a little mercury is put behind the riffles, so as to assist in retaining the gold, and occasionally the riffle-box is supplemented by a series of blankets, which are useful for catching the fine gold.

The tom is cleaned out periodically, and the gold and amalgam are panned. The tom employs two to four men according to the character of the dirt and the supply of water. It is applicable to ground where the gold is coarse, it being incapable of saving all fine gold, of which at least 10% may be lost. The amalgam is retorted.

In Alaska and other remote districts, primitive methods of saving gold are employed in partly developed placer districts. Miners are obliged to make much out of little. With picks, shovels, whipsaws, and canvas hose, which they are able to carry to their claims, they build ditches, flumes, and sluice-boxes and install small hydraulic plants. In Alaska, one pan holds 20 pounds, and 150 pans equal 3000 pounds, or a cubic yard. Fifteen pans make one wheel-barrow, and 10 wheel-barrows equal one cubic yard. In Western Australia, where water is scarce and the alluvium is very dry, resort is had to dry-blowing. A wooden frame, 2 feet square, is made of 3 by 4-inch lumber. Fitted or bolted to each inside corner is a flexible piece of board 3 feet high, and on top of these is fixed a shallow inclined box, fitted with removable trays containing riffle-bars, over which is a coarse screen. This is surmounted by a hopper, into which the material is fed. The operator stands on the windward side and rocks it to and fro. The large pieces run off over the screen, while the finer have to pass over the riffles, the gold gravitating to the bottom. The tray is periodically removed, and the contents are placed in a dish, and either dry-blown or panned. Various improvements to the shaker have been introduced, mainly the addition of bellows by which the tailing is more rapidly worked down the riffles. In dry-blowing, the prospector uses two dishes. He fills one with the material, and from about the level of his head would pour the dirt into the other dish placed on the ground. Then he would pick up the second dish and repeat the process. Each operation would cause dust to be blown away, leaving the coarse material and gold, which latter could subsequently be collected.

At Manhattan, Nevada, and Quartzsite, Arizona, where water is scarce, dry-washing machines are used. Their capacity ranges from 2 to 6 cubic yards of gravel daily. The apparatus consists of a wooden framework, to which is attached a coarse screen, hopper, crank and gears, riffle-board, and bellows. Gravel is passed through a screen with ¼-inch holes into a hopper of say ½₀-yard capacity,

from which it passes on to an inclined riffle-board, 10 by
20 inches, which is also a screen surface with wooden
riffles at right angles to its length. Pulsations from the
bellows keep the material in motion. Under the riffle-
board is a muslin cloth, stretched over the air chamber.
As the material passes over the riffle-board, the heavier
particles are detained by the riffles and drop through the
screen onto the cloth, while the waste passes over the end
of the board or is blown away by the air-blast. The
gold is recovered by panning the concentrates. Gravel
or wash should yield over 50 cents per yard to be profitable by this treatment.

Ground Sluicing. As the best pay dirt is sometimes
at the bottom of creeks, it is necessary to remove the overlying gravel. The stream must first be diverted into a
flume built on a trestle or running on one side of the
creek. The creek bottom being free from flowing water,
the large boulders are piled along the banks, those too
heavy to move being broken by hammers and powder.
A narrow channel built of boulders is thus formed, which
serves to confine the stream and increase its velocity, so
that on being turned back into the creek-bed it is able to
carry off material that it could not otherwise have moved.
Miners often enter this swift-flowing stream, and with
shovels help the larger rocks down stream and off the
claim. From time to time the water is diverted into the
flume, so that the large boulders may be thrown out or
broken up. Where the creek-bed is wide, this temporary
channel must be moved step by step, from one side of
the creek to the other, then perhaps back again in case
the gravel is deep. In this way the mass of gravel is disintegrated and washed away, and the gold is concentrated

in a shallow spot on bedrock. To clean up the bedrock, sluices are laid, beginning at the lowest point on the claim, and into these the enriched gravels are worked by means of wing-dams and shoveling.

Sluices. These consist of a series of troughs formed by planks nailed together. Each sluice-box is 12 feet long, tapering from a width of 18 inches at the upper to 14 inches at the lower end, thus allowing the boxes to fit into one another, and a sluice of considerable length to be formed. Several kinds of riffles are used. An ordinary form is made by fitting round blocks 4 inches thick, sawed from logs a foot or more in diameter, into the boxes; another style consists of poles placed a half inch apart lengthwise in the bottom of the sluice-box; while a third is made of sawed strips of wood, 2 inches thick, and 3 inches wide, placed crosswise, and set at an angle with the bottom of the box, so as to overhang on the upstream side. All of the riffles are held in place by wooden wedges, so that they can be removed for the clean-up, which begins with the upper set of riffles, the concentrate finally collecting in the lowest box.

Hydraulicking or hydraulic mining. In localities where the gulches are deep, the fall of the ground rapid, and the auriferous deposits of considerable thickness, banks of gravel are sometimes attacked by jets of water under high pressure, and the earth washed down and carried through the sluices without being touched by hand. This is called hydraulicking or hydraulic mining. This method is very effective, and under favorable circumstances, such as a plentiful supply of water with good fall, a small amount of gold to the cubic yard (1½ tons) is profitable. One cent a yard has paid in New Zealand,

and 2 to 4 cents in Australia and America. The water
is brought in flumes or pipes to the point where it is required, thence in riveted steel pipes gradually reduced in
size and ending in a nozzle somewhat like that of a fire-
man's hose. This is called a giant or monitor. Figs. 48
and 49 show the arrangement. Fig. 48 shows the monitor

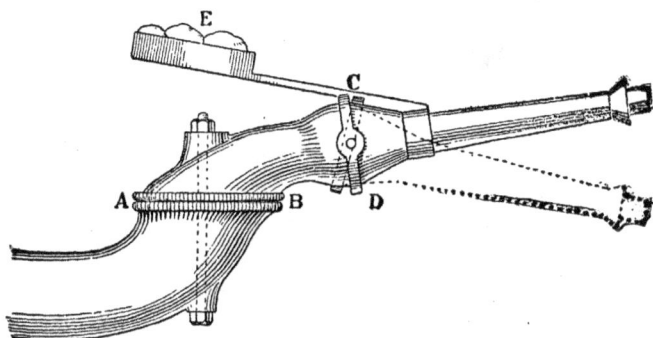

FIG. 48.

movable at *A B* in an ascending, and at *C D* in an inclined, direction. This is done by means of a universal
joint. *E* is a lever loaded with weights, which facilitates
the adjustment of the monitor by the operator in any
direction. The method of operating the arrangement will
be seen in Fig. 49. *A* is the water-distributor or pen-
stock, *B* the nozzle, *C* channels for carrying off the debris
detached from the bank; *D,* piles of larger pieces of rock
which are finally comminuted. *T* is a tunnel through
which the water reaches the sluice, provided with the grating *F* through which the finer gravel falls into the shallow settling-basin *E,* and is distributed by blocks *G*. The

main flow of water and coarser material passes over the
grating F into the main sluice in which the grating H
retains the larger pieces, which are then thrown out at J.
The basins E and the main sluice are paved with wooden
blocks, stones, or steel rails, between which mercury is

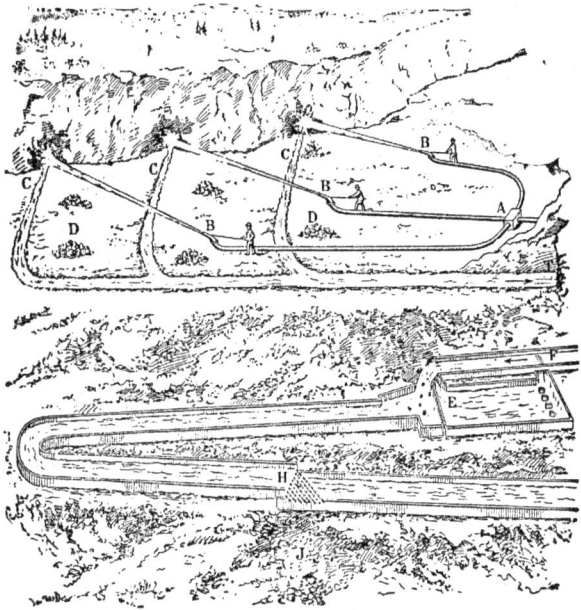

Fig. 49.

placed.   The amalgam recovered is cleaned in a mercury
bath, pressed through stout cloth, and distilled.

One of the great difficulties in hydraulic mining is the
disposal of the tailing, which may amount to several thousand cubic yards daily from a single mine.

Dredging. This method of recovering gold — and platinum and tin — from river beds, bars, and alluvial flats, has grown tremendously since the first apparatus was used in New Zealand in the 60's. It is a cheap system, and gravel as low as 5 cents a yard has paid, although the average value in, say, California is about 10 cents. A dredge is a flat-bottomed boat, with wood or steel hull, fitted with buckets for digging the gravel, washing and screening apparatus, after which the fine material passes over inclined tables to catch the gold, after which the tailing is discharged overboard at the stern. A dredge may thus work its way up stream, or may proceed across a flat plain floating in a pond, cutting out the bank in front of it and piling up the tailing behind. A continuous supply of clear water is necessary for washing when in a pond, as muddy water interferes with saving the gold. Sometimes suction-pumps or grab-bucket dredges are used to lift the gravel, but ladder-bucket dredges are the general type. Some of the 16-cu. ft. bucket boats will dig over 6000 yards daily.

In nearly all alluvial goldfields, whether shallow placers or deep leads, is found a stratum of ferruginous conglomerate, composed principally of rounded and angular fragments of quartz of all sizes, cemented together by the oxide of iron with which the mass is impregnated, and often so hard as to resist everything but blasting. This cement, as it is called, overlies the bedrock, in some places resting on it, in others several inches or even feet above it. In thickness it fluctuates, from 6 inches to 8 feet or more, and its character varies little. It is often highly auriferous, and is worthy of special attention. It should be crushed to a fine powder and tested.

Many particles of fine gold, notwithstanding their greater specific gravity, exhibit the tendency to float in water when undergoing a washing process. To save this fine flour or float-gold, as it is called, experiments have shown that by heating the water to boiling-point or nearly so, these floating particles of gold will subside to the bottom of the pan or vessel.

For lode prospecting a pestle and mortar should be carried. The handiest for traveling is a mortar made from a mercury bottle cut in half, and a wrought-iron pestle with a hardened face, weighing say 5 pounds. A piece of drill-steel serves well. To get the rock fine enough, a screen is required, and the best for the prospector who is often on the move, is made from a piece of cheese cloth stretched over a small hoop. It is often desirable to heat the rock before crushing, as it is thus more easily triturated and will reveal all its gold. Having crushed the gangue to a fine powder, proceed to pan it off in the same manner as washing out alluvial earth, except that in prospecting quartz one has to be much more particular, as the gold is usually finer. Take the pan in both hands, and admit enough water to cover the pulverized rock a few inches. The whole is then swirled around and the dirty water poured off from time to time until the residue is clean quartz sand and heavy minerals. Then the pan is gently tipped and a side-to-side motion given to it, thus causing the heavier contents to settle down in the corner. Next, the water is carefully lapped in over the side, the pan being now tilted at a greater angle until the lighter particles are all washed away. The pan is then once more righted and a little water is passed over the heavy mineral by a shaking motion, when the gold will be revealed in a streak or tail along the bottom. In this operation, as in all others, only practice will make perfect, and a few practical lessons are worth pages of written instruction.

An amalgamation assay that will show the amount of gold in a ton of ore is useful. Take a number of samples; drilling from bore-holes make the best test. When finely triturated, weigh out one or two pounds, place in a black iron pan (it must not be tinned) with 5 oz. of mercury, 4 oz. of common salt, 4 oz. of soda, and about half

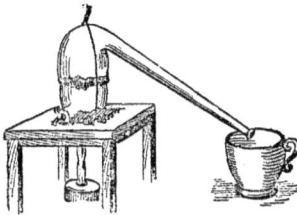

FIG. 50.

a gallon of boiling water. Stir the pulp constantly with a stick, occasionally swirling the dish as in panning off, until it is certain that every particle of the gangue has come in contact with the mercury. Then carefully pan off into another dish so as to lose no mercury. Having got the amalgam clean, squeeze it through a piece of chamois leather, though a good quality of new calico previously wetted will do as well. The resulting pill or ball of hard amalgam can then be wrapped in a piece of brown paper, placed on an old shovel, and the mercury driven off over a hot fire. A clay tobacco pipe, the mouth being stopped with clay, makes a good retort. To make such a retort, Fig. 50, take two new tobacco pipes similar in shape, with the biggest bowls and longest stems procurable. Break off the stem of one close to the bowl and fill the hole with well-worked clay. Set the stemless pipe on end in a clay bed, and fill with amalgam, pass a piece of thin iron or copper wire beneath it, and bend the end of the wire upwards. Now fit the whole pipe, bowl inverted, on the under one, luting the edges well with clay. Twist the wire over the top with a pair of pliers until the two bowls are fitted closely together, resulting in a retort that will stand any heat necessary to distill mercury thoroughly. The residue, after the mercury has been driven off, will be retorted gold, which, on being weighed and the result multiplied by 2000 for 1-pound assay, or by 1000 for two pounds, will give the amount of gold per ton that an ordinary stamp-mill might be expected to save. Thus 1 grain to the pound, 2000 pounds to the ton, would show that the ore contained 4 ounces, 3 pennyweights, and 8 grains per ton.

Darton's gold test. A number of methods have been proposed to detect the minute quantities of gold in rocks, and the following is one that requires little time and is reliable:

Small pieces are chipped from all sides of a piece of rock, amounting in all to about ¼ oz. This is finely powdered in a steel mortar, and well mixed. About half of it is placed in a large test-tube, and then partly filled with a solution made by dissolving 20 grains of iodine and 30 grains of iodide of potassium in about 1½ oz. of water.

The mixture thus formed is thoroughly agitated by shaking and warming. After all particles have subsided, dip in a piece of pure white filter-paper, allow it to remain for a moment, then let it drain, and dry over the spirit lamp. The paper is then placed on a piece of platinum foil held by pincers, and heated to redness over the flame. The paper is quickly consumed, and after heating further to burn off all carbon, is allowed to cool, and then examined. If at all purple, gold is present in the ore, and the relative amount may be approximately deduced as much, fair, little, or none. There is no compound that would be formed from natural products by this method which would mislead by staining the ash to a color at all similar to the distinctive purple of finely-divided gold.

A variation of this test is given by Thorpe and Muir in ' Qualitative Chemical Analysis ' as follows:

Five or ten grains of the finely powdered mineral are shaken with alcoholic tincture of iodine, prepared by dissolving ½ oz. of iodine and ¼ oz. of iodide of potassium in 1 pint of rectified spirits.

The insoluble matter is allowed to settle, a piece of Swedish filter-paper is dipped into the solution and incinerated after drying. If the ash be purple in color, gold is present. To confirm the presence of gold, treat the ash with a few drops of aqua regia, evaporate to dryness at a gentle heat, and dissolve the residue in water. Pour this solution into a beaker, which is set upon a sheet of white paper. A solution is now prepared by adding ferric chloride to stannous chloride until a permanent yellow color is produced. This solution is diluted, a glass rod is dipped into it, then into the gold solution. A bluish-purple streak in the track of the rod confirms the presence of gold.

Geology of Gold. Native gold is found, when in

place, with comparatively few exceptions, in the quartz veins that intersect metamorphic rocks, and to some extent in the wall-rock of these veins. The metamorphic rocks thus intersected are mostly chloritic, talcose and argillaceous schists of dull green, dark grey, and other colors; also much less commonly mica and hornblende schist, gneiss, diorite, porphyry, and still more rarely granite. A laminated quartzite called 'itacolumite' is common in many gold regions, and sometimes specular schists or slaty rocks, containing much foliated specular iron (hematite) or magnetite in grains.

Gold exists in quartz in strings, scales, plates, and in masses, which are sometimes an agglomeration of crystals. The scales are often invisible to the naked eye, massive quartz that apparently contains no gold frequently yielding a considerable amount by assay. Gold is always irregularly distributed, and never in continuous pure bands of metal like many metallic ores. It occurs both disseminated through the mass of quartz and in its cavities.

In studying the geologic aspects of this subject, and making a practical application of one's knowledge to the search, it may be stated that the original source of gold must have been at great depths, but in this book it is not necessary to discuss the various theories of ascending and descending waters and lateral secretion. Suffice it to say that quartz veins and gold were precipitated from solutions which eventually filled the fissures called veins.

Cellular quartz has been found frequently with gold within the cells, the material that surrounded the gold having become decomposed, thus releasing the undecomposed gold. Miners often judge the character of rock by surficial study. Quartz that is dull white in color, uncrystallized, and shows no traces of brownstone (decomposed pyrite) is called ' hungry,' and not expected to yield good returns. On the other hand, quartz that is well honeycombed from decomposition of pyrite, stained brown on joints and faces, and shows decomposed pyrite between the crystals, promises well.

Enrichment of the upper or surficial parts of a gold-quartz vein is often the result of simple concentration. The leaching and removal of the pyrite is effected without the shifting of the gold, which remains behind in a honeycombed mass of iron-stained quartz, making specimen ore. Thus this part of the vein loses weight without reduction of volume, so that the ore is so much richer per ton. In some cases the leaching of the upper portion of a gold vein may produce solutions in which the gold is soluble sufficiently to cause it to migrate downward and to be precipitated near the water-level, which not infrequently coincides with the base of the zone of oxidation.

Gold, therefore, is to be expected and looked for in granitic regions (Fig. 51), and in those rocks and from those gravels and sands that owe their origin to such regions. The auriferous belt of California extends along the lower slopes of the Sierra Nevada, which is formed of granites, flanked by crystalline schists and other rocks of the Jurassic. It consists of quartz veins striking in the same directions as the beds, and containing numerous metallic sulphides which all carry gold. It requires much judgment, general exploration, and knowledge of the region before the prospector can expect to find gold, or before he should begin the search. With a full knowledge of the geologic conditions of the country and acting in accordance with the above facts, the prospector will soon come upon traces of gold, if any exist.

In looking for indications, the prospector should never pass an ironstone ' blow-out ' without examination. Such an outcrop may cover a gold, silver, copper, or tin lode. Besides the general instructions given above, consider-

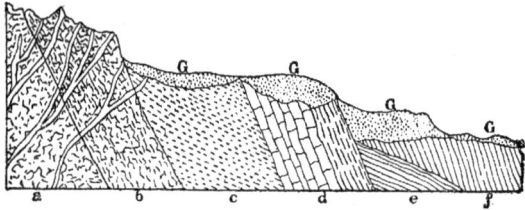

FIG. 51.— SECTION SHOWING THE TWO CONDITIONS UNDER WHICH GOLD IS USUALLY FOUND IN ROCK AND DRIFT.

a. Granitic and gneissic rocks penetrated by greenstones and porphyrytic rocks containing gold finely disseminated. *b.* Micaceous, talcose, and argillaceous slaty rocks, supposed to be Laurentian and Cambrian. *c.* Silurian and Devonian strata. *d.* Carboniferous, limestone and grits. *e.* Coal measures. *f.* Permian and newer rocks. *G, G, G, G.* Drift, filling hollows in rocks with gold, especially at the base of the drift.

able study should be devoted to the peculiar and seemingly irregular deposits of gold where it does not appear to have been washed down from any higher levels. In California and some other districts, free gold has been found in drifts and sand and in the beds of streams which have not only been filled up, but have been buried under sandstone or other rocks, but the whole country has apparently been raised, or the surrounding region has sunk so as not to show any very considerable elevation beyond where the gold deposits have been formed. But even in this case, the general rule has been shown to be correct, for these deposits have been proved to be in the beds or channels of ancient rivers, which had either been dried up and overflowed by eruptions of lava or basalt, and again by floods bringing new soil and creating sedimentary rock; or the country has been raised, or subsidence of a great extent of land has taken place. In many cases, however, no subsidence has occurred, but only overflow and filling up through ages, and the actual sources still remain elevated.

Such events as those just described do not happen without leaving traces, which, to the eye of a skilful prospector, are evidences of some such movements and changes. He will then proceed to make a successful opening only after carefully examining a large tract, as it is from extended survey that he may judge better the relation of surficial parts to the greater depths of even small areas.

Those rocks that lie more immediately over granite, and which, although they owe their origin to a sedimentary condition, have been subjected to heat and heated waters, as is supposed, are called 'metamorphic rocks.' They were probably first formed from disintegration of the most ancient rocks, and have brought with them fragments of gold. These metamorphic rocks have been changed from ordinary sedimentary rock by the action of heat and pressure, and the influence of such treatment may be suspected by their appearance being crystalline; that is, the fine grains that compose them, as well as the larger grains, are angular, whereas the materials of purely sedimentary rocks are fine without angular shape. The

larger part of granite is supposed to have been metamorphic or changed, as the word means, or altered
merely by the action of heat into a crystalline form or
mass.

The igneous rocks are those whose forms are due to
having been melted and driven to the surface through
fissures in the overlying rocks. They are composed of
feldspar, hornblende, a little quartz, with comparatively
small proportions of other substances, and are called by
various names according to the composition. The metamorphic granite contains quartz, feldspar, and mica; the
igneous granite contains little or no quartz. Syenite-
granite contains hornblende in place of mica. Sometimes the mica is very black, as hornblende is; and in that
case may be distinguished from the latter by its more easy
cleavage, as has been shown, under a sharp knife-blade.
This black mica is described under biotite. There is a
syenite containing no quartz called hyposyenite. These
rocks are not the original source of gold, but at present it
is largely in these metamorphic rocks that the most profitable gold is to be found, more especially in the quartz
veins which have intersected these rocks. One, therefore,
of the most important studies of the prospector is to ac-
quaint himself with the appearance, positions, and de-
partures of these metamorphic rocks. In many places
where the alluvial gold, derived from the gold-bearing
gravels, has almost ceased to be worth working, there still
remain sources undiscovered. These may probably be
traced back to some outcrop or to some ancient elevation
having subsided.

The above remarks are applicable to exploration for
other metallic ores than gold. They apply to silver, espe-

cially to tin ores, and with some modifications, to copper and to quicksilver ores, as will be shown.

Areas near where there are contacts of granite and greenstone (diorite) should be examined. Pieces of quartz, jasper bars, and sometimes the outcrop of a quartz lode will influence prospectors in a close search. If float ironstone, quartz, or lode matter be found, they are crushed and panned. If gold-bearing, the general direction in which they have traveled is carefully and methodically followed. As soon as any traces of gold are found, the prospector, if he be in country where there is much alluvium, may try the 'loaming' method, which originated in Western Australia, and was the means in mid-1919 of discovering an important deposit.

As described by C. M. Harris, who is largely responsible for the new method, a loam bag, 6 feet long and 6 inches diam., made of unbleached calico, with tapes sown to it every 9 inches from top to bottom, is part of the outfit. If the ground be on a rise, the prospector starts on the fall or where he found the floaters or indications, makes a hole 4 to 6 inches deep, takes a sample from the bottom, and puts it in his loam bag. He then holds the bag perpendicularly and shakes it well so that all the fine material falls to the bottom, then No. 1 tape is tied. He next proceeds for about 4 yards across the country and repeats the operation until he has 8 samples, and 8 holes in a row. Having marked the spot he proceeds to pan the samples, carefully noting and driving a long stick into the ground in each hole in which the tail or prospect (if any) of gold was obtained. If there are any colors of gold, he takes another parallel course 10 yards ahead, starting in front of the first sample which showed the tail, and proceeding at shorter distances than in the first line, until opposite the samples last taken in the first line, which showed a tail.

When this second row of samples has been panned, they will determine whether gold continues further; if it does, or has a tendency to spread out, then a wider line must be taken ahead again, and so on until the loam gold cuts out. The prospector then starts to costeen (trench), still proceeding up-hill, and the debris should be panned right along to determine (should no lode or vein be visible) if the gold is going down in the cement (a ferruginous limey conglomerate) or the sub-soil. If this be so, the prospector knows he can go on sinking, as sometimes he may have to do for several feet below the surface, often without the slightest sign of the lode until it is cut. If he strikes a long shoot of gold, the loams containing a tail may be widely spread. On the contrary, the usual characteristic is that the tail in the loams will gradually narrow until they cut out. When satisfied as to the probable strike of the hidden outcrop, a costeen or small shaft is sunk through the overburden, and then a crosscut to the lode. Constant sampling and panning are necessary. Loaming is more reliable than trenching, and is also quicker and cheaper. If water is not available for panning, dry-blowing, as described on page 127, should be employed. These are the main features of the system, which is modified according to circumstances.

Gold in combination. Native gold is not the only condition in which this metal is found. The combination of gold with various oxides and sulphides of other metals are valuable, and should be studied.

In almost all gold-bearing regions iron sulphides carry

gold, and in some gold is found only in this mineral. Hence it is well for the prospector to determine the presence of gold in the pyrite or whatever sulphide may be found.

1. To separate gold in metallic sulphides, such as iron pyrite: Powder the sulphide as fine as possible. Put an ounce into a clay crucible, and heat to a low red heat for an hour, or until there is little escape of sulphur fumes. Remove the crucible and put its contents into a porcelain dish. Pour over the roasted powder three fluid ounces of strong nitric acid, by drops, until all violent action ceases. Add water, 8 or 10 fluid ounces. The gold, if any, will appear as a fine, black powder. Filter and dry, pick out a small particle of the powder and mash it upon a hard surface, iron or agate, in an agate mortar; if it is gold, it will show the gold color. A sufficient quantity of the dried powder may be placed upon a piece of charcoal, and by means of either the outer or inner flame of the blowpipe it may be melted, and both by its color and softness proved to be gold.

2. Another method of detecting and separating the gold, where the above cannot be used, is by pulverizing the sulphide ore very fine, and mixing it with three or four times its weight of caustic potash or caustic soda, then subjecting the crucible, which contains the mixture, to a low red heat till all the contents cease agitation and become quiet. Remove the crucible, wait till all is cool, and then add hydrochloric acid equal in amount to three or four times the bulk of the mass. To this, after standing three or four hours in a warm place, add the usual nitric acid (about an ounce), after transferring all the liquid to a porcelain dish, or, next best, to a beaker-glass.

Let it stand in a warm place for an hour, then add a little more nitric acid (about half an ounce), stir well with a glass rod or strip of glass, and let stand again for an hour or two. Examine carefully, and if it seems to have been dissolved more thoroughly than before, add a little more nitric and warm again, stirring well as before. If no more seems to be dissolved, then filter and wash the sediment in the filter and let it dry, and remove the filter and contents for further examination. Next precipitate the gold from the filtrate by pouring into it a solution of ferrous sulphate. (Any clear green crystals of copperas — sulphate of iron — from a drug-store, dissolved in clear rain water to saturation point, filtered, and kept in corked bottles, will answer this purpose.) Let the solution stand in a warm place for an hour, add a few more drops, and if any further precipitation takes place, add half an ounce of the sulphate, stir it again, let it remain an hour longer in a warm place till all precipitation ceases. Decant the supernatant clear water and carefully transfer the remainder to a filter-paper, a little at a time, to avoid breaking the paper. Then rinse the porcelain dish to get all particles upon the filter, and when all the liquid has passed through, let it dry, remove all the contents of the paper to a small porcelain capsule or crucible, and apply the heat of the blowpipe to burn off the paper or any organic substance which may have got into the powder — the gold remains, and may be gathered upon charcoal and melted into a globule by the concentrated flame of the blowpipe, if in small quantity. Lastly, examine the contents of the filter that was laid aside, and, if there is any appearance of gold separate it under examination by a lens.

It must be remembered that the above process does not extract all the gold from a pyritic ore unless conducted with more care and time than here suggested, but it is sufficient to reveal the fact that the ore is valuable, also that gold is only mechanically mixed with sulphides. A little practice will enable the operator to become expert in these methods. A great deal more depends upon one's skill than upon the cost of the appliances.

Rule for finding the amount of gold in a piece of auriferous quartz:

The specific gravity of gold is 19.000, and that of quartz is 2.600.

(These numbers are given merely for convenience in explaining the rule; they do not accurately represent the specific gravities of all quartz and quartz gold.)

1. Ascertain the specific gravity of the ore. Suppose it to be 8.067.

2. Deduct the specific gravity of the ore from the specific gravity of the gold; the difference is the ratio of the quartz by volume: $19.000 - 8.067 = 10.933$.

3. Deduct the specific gravity of the quartz from the specific gravity of the ore; the difference is the ratio of the gold by volume: $8.067 - 2.600 = 5.467$.

4. Add these ratios together and proceed by the rule of proportion. The product is the percentage of gold by bulk: $10.933 + 5.467 = 16.400$. Then, as 16.400 is to 5.467, so is 100 to 33.35.

5. Multiply the percentage of gold in bulk by its specific gravity. The product is the ratio of the gold in the ore by weight: $33.35 \times 19.00 = 643.65$.

6. Multiply the percentage of quartz by bulk (which must be 66.65, since that of gold is 33.35) by its specific

gravity. The product is the ratio of the quartz in the ore by weight: $66.65 \times 2.60 = 173.29$.

7. To find the percentage, add these two ratios together and proceed by the rule of proportion: $633.65 + 173.29 = 806.94$. Then, as 806.94 is to 633.65, so is 100 to 78.53. Hence, a sample of auriferous quartz having a specific gravity of 8.067, contains 78.53% of gold by weight.

Following are brief descriptions of the geologic features of certain gold deposits in various regions; a little study may result in their being suggestive in field work:

At the Pato and Nechi properties in Colombia, South America, which are being worked by dredges, the area lies in an oval basin surrounded by crystalline rocks, cut through by the Nechi river. Within the basin are flat tables of gravel deposits whose edges form terraces. Bedrock of each bench is clay, in which are beds of peat or brown coal. Each bench is a re-concentration of the bench above it, and represents a former stationary level of the river. Gold in the gravel was originally brought in from the surrounding hills, all of which are auriferous, by the small storm streams. Successive concentration has resulted in the present pay-gravel. In each concentration, part of the finer-grained gold floated away, so that the average grain of the lower deposits is fairly coarse, although nuggets or rounded grains are seldom found. The gold is of fairly uniform sized thin flakes.

Colombia is attracting considerable attention by reason of its possibilities for gold and platinum in alluvial deposits. That country is looked upon as the future supply of platinum.

The Juneau gold belt of southeastern Alaska can be traced for 50 miles northwest and 40 miles southeast of

Juneau. It consists of a single band, several hundred feet wide, in which stringers and veins of quartz, carrying gold, occur in a slate formation near its contact with some altered volcanic rock known as greenstone. In this belt are the lodes worked by the Alaska Gastineau and Alaska Juneau companies, which extract about $1 per ton. Gold is usually free, and is associated with pyrite, galena, blende, and arseno-pyrite.

The Mother Lode of California is a mineralized belt about 150 miles long and several miles wide in the foothill section of the Sierra Nevada. In it are a number of quartz veins, which are either in slate or at the contact between it and greenstone dikes. Calcite and dolomite also form part of the gangue rock. Gold occurs free and associated with 1 to 3% of sulphides — pyrite, arseno-pyrite, galena, tellurides, and copper minerals. The ore is not rich, yielding from $3 to $7 per ton. The veins are persistent to below 4000 feet. Treatment of the ore is simple.

There is doubt as to the origin of the 'reefs' on the Witwatersrand of the Transvaal, South Africa. These contribute 40% of the world's gold output. The ore is a conglomerate, consisting of quartz pebbles cemented by grains of sand and pyrite, the cementing material carrying the gold, which averages 6 dwt. per ton. The enclosing rock is quartzite, which lies within a region of igneous rocks such as granite, basalt, and diabase.

At Goldfield, Nevada, the most important rock is dacite. In this are irregular ore-shoots consisting of quartz, pyrite, telluride, and gold.

Pocket mining has yielded some rich gold ore, especially in Siskiyou county, California. Search is made for calcite, a replacement in slate. The gold-bearing belt of the county consists of metamorphic slates, granites, diorites and limestones, with occasional intrusive masses of porphyry, trap, and syenite. In some places the belt is veined and seamed with stringers of quartz, rich in gold, designated in hydraulic mining as 'seam diggings,' and easily worked with a stream of water under heavy pressure. In other places, where the formation has been fissured, or near the line of contact quartz veins occur singly or in groups, and are generally small but rich in gold. The following are examples of pocket mines: (1) ore in diabase; (2) ore in slate; (3) ore in limestone and diorite; (4) ore in hornblende and grano-diorite; (5) ore in granite and schist; and (6) ore in hornblende schist. All of these have produced a good deal of gold, some as much as $10,000 from a pocket.

At the Aztec mine at Baldy, New Mexico, is another instance of pockets. There, gold ore is often found at the contact of shale and quartz-monzonite porphyry. Folding has fractured the shale at the crests of the folds, opening minute fissures, many of which are filled with calcite. This calcite carries pyrite, chalcopyrite, and sphalerite, which seem to coat the gold so that it does not appear metallic. A good deal of the ore came from the lower part of the conglomeratic sandstone, or from fractured portions of the shale close to the contact. Over 2100 tons averaged $107.60 per ton from this mine.

The ore-zone in the Mt. Champion gold mine at Leadville, Colorado, is a granite intrusion in gneiss, accompanied by an alaskite dike with several branches passing through the granite, the latter being mineralized and faulted.

Lode material of the Unsan mines in Korea, which have produced over $30,000,000 of gold, is quartz containing pyrite, galena, sphalerite, gold, and a little silver. A predominance of galena is generally accompanied by an increase in gold. Free gold is rarely visible, and the ore is only worth about $6 per ton. Graphite is found in small quantities. Coarsely-grained grano-diorite is the important country rock. Dikes and faults are common.

In Sarawak (northwestern Borneo, Dutch East Indies) is a peculiar gold deposit. Overlying limestone are thin beds of marl, which in turn is overlain by a series of shales, sandstones, grits, and conglomerates. Faulting and intrusions by quartz-porphyry dikes are common. The ore is mined by open-cuts, and treatment is simple, just crushing through ¼-inch screens and cyanidation in open vats. The average gold-content is under $4 per ton.

In the Zortman district of Montana are gold-bearing porphyry dikes worth about $3 per ton. Three companies are treating this class of ore by simple cyanidation.

Ontario, Canada, is a region of great possibilities for new discoveries, especially gold. Outcrops are generally prominent, and it was from them that such centers as Porcupine developed. The Dome was a particularly striking outcrop. This field is producing 400,000 oz. of gold annually, and will increase. The gold is in pyritic gold-quartz lodes in pre-Cambrian schists. Diabase dikes cut the lodes. Faults are common, and need much study.

Other districts in Ontario are Kirkland Lake, Boston Creek, and Matachewan. Prospecting in northern Manitoba and Quebec is on the increase, and gives encouraging results.

In the great Kolar goldfield of India, where the main lode is being opened below 4000 feet depth, the ore is in a hornblende-schist, which lies in a belt of granites. The lode is extremely folded in spots. The ore is quartz carrying a small amount of sulphides, and over 6,000,000 tons averaged 1 ounce per ton.

At Cripple Creek, Colorado, quartz is the most important vein-forming mineral. The ore is a sulpho-telluride. The geology is complex, as the ore-zones are in vent of an old volcano. Briefly, the veins are fissures in breccia and granite. The Independence vein had a prominent quartz outcrop in the granite (silica 72%), but no outcrop in the andesite (silica 59%).

At Kalgoorlie, Western Australia, the productive rocks — diorite or greenstone — are within a belt flanked by granite and gneiss. The lodes, which have no defined walls, consist of schistose matter which is highly silicious, carrying pyrite, tellurides, and free gold.

Limestone formations should generally be avoided when prospecting for gold, although at Manhattan, Nevada, the White Caps mine has a deposit in a replacement of limestone by quartz, pyrite, arseno-pyrite, and stibnite, while another has contact and chamber deposits in limeshale. Calcite is abundant, and faulting is common.

The Nickel Plate gold mine in British Columbia contains ore deposits of contact metamorphic origin, which occur at the contact of dikes and sheets of gabbro in altered limestone. Arseno-pyrite is the principal sulphide mineral.

In the McIntyre-Porcupine mine in Ontario the best gold deposits are in an altered volcanic schist, at or near the contact with the quartz-porphyry. The ore-shoots consist of quartz more or less inter-banded with mineralized schist. In the Fortuna mine in Arizona, the quartz vein is in schist, near gneiss and granite masses. The outcrop shows iron and copper oxides, but the ore is free-milling. At Gympie, Australia, the gold veins lie in sedimentary rocks, such as sandstone, shale, conglomerate, limestone, and breccias. The important beds are the so-called 'slates,' which are shales, sandstone, and graywacke, and it is only where the lodes intersect these that they are auriferous. These beds contain calcite and graphite, an abundance of the latter often resulting in local enrichment. At Central City, Colorado, the country-rock is pre-Cambrian schist and gneiss. Wide zones of mica-schist are recognized as unfavorable for extensive quartz lodes, while the more silicious gneiss is recognized as favorable.

SILVER

Native silver exists in various shapes, as in small grains in ore, as wire, tree-like, in small octahedral crystals, and in other forms. Color and streak, silver-white; when found in veins is usually tarnished dark brown on the surface. Hardness 2.3 to 3; specific gravity 10.1 to 11.1, according to purity. It is never found absolutely pure, but contains some gold, frequently a little copper. It is always sectile and malleable, and in this respect easily distinguished from a mineral frequently mistaken for native silver, namely, mispickel (arseno-pyrite), which is an arsenide of iron, having the appearance of silver, but is always brittle. Native silver is generally present in gold, lead, and copper districts.

Before the blowpipe, on charcoal, native silver is identified from tin, zinc, antimony, or bismuth, by the fact that it melts and leaves no whiteness or any other appearance of oxide upon the coal around the globule. Tin will leave a white film, lead a yellow, and zinc a yellow, which whitens on cooling.

A chemical test for silver is to dissolve the metal in nitric acid in a test-tube, preferably with the heat of an alcohol flame, but not to boiling point. Add an equal amount of pure water (clear rain water will do), then add in several drops of a solution of common table salt or muriatic acid. If a cloudy white precipitate forms, which settles and blackens after exposure of a few minutes to daylight, the substance is silver.

It should be remembered that this test is for silver alone, since lead and mercury are also precipitated as a white cloud by the same solution, but neither blackens by exposure to light. This distinguishes silver. If, however, further proof is needed, add strong liquid ammonia to the test-tube; the precipitate is dissolved if it is that of silver; it is not if it be lead, and it is blackened by the ammonia if mercury.

If there is much copper in the silver it may be detected by dipping a clean strip of polished iron or steel into the solution, when the metallic copper will immediately appear upon the surface of the iron.

It must not always be supposed that native silver is metallic or white in appearance, for it is readily tarnished by sulphur, and proximity of sulphur in other minerals or in water may greatly discolor the silver.

Comparatively speaking, little of the metal mined is derived from native silver. Most of the silver of commerce is obtained from some of the minerals named below, which are combinations of silver with other metals, and with sulphur or chlorine, in which condition they bear no resemblance to native silver.

In all silver minerals of any commercial value, the tests given are usually sufficient to detect its presence.

Argentite or silver glance. This is a sulphide of silver, and is found in masses, but when crystallized it occurs in cubes or octahedral forms. When freshly broken it has a metallic luster, otherwise it is of a dull gray or lead appearance. It is soft and highly sectile, and its streak or the color of its powder is the same as that of the mineral itself — lead-gray to black — and rather shining. Chemical composition is silver, 87%; and sulphur, 13%. Its hardness is 2 to 2.5, and specific gravity, 7.1 to 7.4.

The ore is soluble in nitric acid, and on adding common salt to the solution, a white precipitate is thrown down, which blackens on exposure to sunlight. It is very fusible at the temperature of an ordinary flame, giving off an odor of sulphur when heated. Before the blowpipe on charcoal, with or without carbonate of soda, it yields a white globule of metallic silver, which can be flattened under the hammer.

The ore in an amorphous state is most common in earthy vein-stuff (called metal azul) or with pyritic minerals, especially galena. It is rarely recognizable by form or physical character, as rich quartz only differs from ordinary by its pale bluish-gray tint, and argentiferous galena is, as a rule, undistinguishable by sight from that containing no silver.

Argentite is a primary silver mineral in many silver districts, and is usually found with cerargyrite, stephanite, polybasite, and pyrargyrite. Argentite is one of the principal minerals at Tonopah, Nevada.

Cerargyrite or horn silver. This mineral is a chloride of silver associated with other ores of silver, usually only in the upper parts of veins, also with ochreous brown iron ore, and with several copper ores. Its formation is due to solutions carrying alkaline chlorides, obtained from the overlying strata, acting on the silver minerals of veins and forming solutions of silver chloride, from which the mineral is precipitated in fissures or cavities of the country rock. Its luster is waxy and resinous; fracture is conchoidal; color is greenish-white, pearl-gray, brownish, dirty green, and on exposure brownish or purplish. It yields a gray, shining streak. It is translucent on the extreme edges, and has a waxy appearance. It cuts like horn or wax, and on an outcrop looks like dirty cement. It contains 75.3% silver, and 24.7% chlorine when unmixed or nearly pure.

A polished piece of iron may be slightly coated with silver if a piece of horn silver is moistened and rubbed upon the iron.

Horn silver is easily fusible, melting in the flame of a candle. Heated with carbonate of soda on charcoal, it yields a globule of metallic silver.

This mineral, in various degrees of impurity, forms a

large part of the silver-bearing ores of some mines in South America, at Shafter in Texas, Rochester in Nevada, and at Broken Hill, Australia.

Stephanite or Brittle Silver ore is a silver sulphide with antimony, and is found in masses and sometimes in rhombic prism crystals in veins with other silver ores. It is easily distinguished from silver sulphide by the fact that it is brittle, while the glance, if fairly pure, may be cut with a knife into chips without breaking.

This ore is black or iron-gray, has a hardness of 2 to 2.5, and a specific gravity of 6.2 to 6.3, and when pure contains 68.5% of silver, the rest being antimony with some other minerals, usually iron or copper. It is an abundant silver ore in the Comstock Lode, in the Reese River, Humboldt, and other regions of Nevada, and at silver mines in Idaho.

On charcoal, under the blowpipe, it decrepitates and coats the coal with a film of antimony, which, after considerable blowing, turns red, and a globule of silver is obtained.

Polybasite. This is another sulph-antimonite of silver, and closely resembles stephanite, and the two are frequently mixed and difficult to distinguish.

Ruby Silver. Several ores of silver contain arsenic and antimony as well as sulphur. The most important of these are the dark-red, sometimes black, mineral called 'pyrargyrite' (the dark ruby silver), which contains 59.8% silver, 17.7% sulphur, and 22.5% antimony; and the light-red silver ore, known as 'proustite' (the light ruby silver), with 65.5% of silver, besides sulphur, and may have a grayish appearance. Proustite has been found in masses of several hundred pounds weight in the

Poorman lode, Idaho. In Mexico it is worked extensively as an ore of silver.

Both these minerals occur massive, granular or as prismatic crystals. They resemble each other closely in their characteristics, their hardness being 2 to 2.5, and the specific gravity of pyrargyrite 5.8, and that of proustite 5.6. Both have a red streak, and an adamantine and sub-metallic luster.

Ruby silver is a secondary silver mineral, and is associated with argentite, polybasite, tetrahedrite, and other silver minerals.

Before the blowpipe, pyrargyrite gives off dense antimony fumes, while proustite yields arsenical fumes easily recognized by their garlic odor. Heated on charcoal with carbonate of soda both minerals give a globule of metallic silver.

Nitric acid extracts the silver from these ores, forming a solution, in which salt throws down a white precipitate, blackening on exposure to sunlight.

Bromyrite or silver bromide. This is a mineral containing bromine 42.6% and silver 57.4%.

Testing silver ores. A simple, but rough, method of testing ores when chlorides are the minerals chiefly worked, is by powdering the ore, mixing it with a solution of hyposulphite of lime, which dissolves the chloride, and then adding sodium sulphide, which forms a dark-colored precipitate if much silver be present. It is evidently impossible to estimate in this way the amount of silver, but it affords a good test whether the ore is of value or not.

Reference may here be made to what are called ' argentiferous ' (silver-bearing) minerals, comprising ores of lead or copper, in which more or less silver is present. These are as follows: galena (sulphide of lead); bournonite (sulpho-antimonide of lead and copper); tetrahedrite (antimonial gray copper); tennantite (arsenical gray copper); mispickel (arseno-sulphide of iron); zinc-blende (sulphide of zinc); and chalcopyrite (copper pyrite).

These minerals are described under their respective heads in this book. When argentiferous, they do not give evidence of the presence of silver, unless they are submitted to the process of assay. A simple test is to grind the ore fine, and mix a few ounces with $\frac{1}{10}$ of its weight of salt, and $\frac{1}{20}$ of copperas. This is placed in an old frying pan and heated gently so long as the smell of burning sulphur can be detected, the mass being stirred continuously with a thin bar of iron. After all the sulphur has been driven off, the heat is increased for a few minutes to a light red, and the mass stirred until it swells up and becomes sticky, care being taken not to fuse the ore. The charge is then taken out and allowed to cool, and after a little more salt has been added, and the ore mixed with water to the consistence of mortar, a strip of sheet copper previously cleaned, is inserted and left there for 10 minutes. The copper is then removed, washed in clean water, and, if any silver is present, it will be coated with a white substance which will be heavier or lighter according to the richness of the ore and, if rich, will appear gray and rough. The pan should be smeared with clay or mud and dried before being used.

Geology of Silver Ores. The most valuable ores occur in the earlier or more ancient rocks, such as the granitic or gneissoid rocks, clay-slates, mica-schists, older limestones, and in the metamorphic rocks.

The following are examples of the deposition of silver ores: In the new Divide district of Nevada, the rocks consist of a series of volcanic flows, breccias, and tuffs, with related intrusions similar to those of the Tonopah series, 6 miles north. The rich ore in the Tonopah Divide mine is in coarse-grained rhyolite-breccia. Above this in turn are a fine-grained rhyolitic tuff, a thin layer of rhyolitic-breccia, finally a fine-grained rhyolite. The outcrop of the vein is stained with red and yellow iron oxides, not very attractive in appearance, neither is the rich ore, the gangue of which is mainly kaolinized breccia.

At the new and rich Dolly Varden mine in British Columbia, the ore is a quartz, carrying pyrite, galena, ruby, silver, and native silver, in andesite country.

At Pachuca, Mexico, a district that has the great Real del Monte and Santa Gertrudis silver mines, the geology is very simple. All veins are fissures in a uniform deep-seated andesite boss. Vein-filling is quartz, with a little calcite. The total mineral content — iron pyrite, galena, and blende.— is only 4 or 5% of the ore.

Vein systems of the Dolores mine in Chihuahua, Mexico, consist of rhyolite dikes in andesite, and the veins are fissures in these dikes. Ore is a hard quartz carrying 2% of sulphide of iron and silver. The value is 40% gold and 60% silver.

At Tonopah, Nevada, the geology is complex. There is a series of volcanic rocks, partly lava flows, partly intrusive sheets and masses of quickly varying thickness overlie one another irregularly. The productive veins are in trachyte (a volcanic rock of alkaline feldspar with black mica and hornblende), overlaid by andesite and intruded by rhyolite. The veins are cut through by the andesite, and do not outcrop as a rule, but terminate against the andesite at depths up to 1200 feet. They are fissure veins that for some distance lie on or in planes of contact between eruptive rocks.

At Rochester, Nevada, the enclosing rock in one mine is a highly altered sericitized rhyolite, varying from a soft and friable talcose or schistose product to an extremely tough silicified variety. Cerargyrite is the principal silver mineral. Sulphides are uncommon. The ratio of gold to silver is 1 to 300. In another mine the deposit is a shear-zone of rhyolite with fissure veins, which carry silver chloride and free gold. In a third mine there are quartz veins in rhyolite, carrying oxide and sulphide ore with silver and gold.

The Waihi mine in New Zealand contains a number (16) of large calcitic quartz veins in rhyolite, andesite, and dacite with argentite as the principal silver mineral. The ratio of gold to silver is about 1 to 7. Most of the lodes do not outcrop.

Quartz veins in the Innai silver mine in Japan traverse all of the rocks, namely, tertiary tuff, shale, and sandstone, as well as liparite and andesite, the two latter intruding the tertiary.

Good silver prospects exist in pre-Cambrian formations in Lemhi county, Idaho. Owyhee county has been a large producer of silicious silver-gold milling ore. At the Demming mine the veins are in granite, which are cut by cross fissures. The ore is a sulphide and is concentrated. As the Ramshorn the veins are in a black altered slate. Quartz and spathic iron are gangue minerals, with streaks of tetrahedrite.

The silver ore of Cobalt, Ontario, occurs in narrow calcite veins, which are in conglomerate and Keewatin (altered eruptives — schists) formation. The shoots are irregular, and those from 1 to 3 inches wide often contain
as much as 3000 oz. per ton. Besides silver, the other
valuable minerals are arsenic, cobalt, and nickel.

Deposits in the Batopilas mine of Chihuahua, Mexico,
are also in calcite, and rich concentrations of silver are
frequently found.

PLATINUM

Color and streak is steel-gray; luster — metallic, bright;
isometric, but is seldom found in crystals; hardness — 4
to 4.5; and specific gravity — 16 to 19. It is as heavy
as gold, and, therefore, easily distinguished and separated
from lighter materials; and is malleable and ductile. Before the blowpipe it is infusible; and is not affected by
borax, except when containing some metal, such as iron
or copper, which gives the reaction. Soluble only in
heated aqua regia (3 parts hydrochloric and 1 part
nitric acid).

Platinum occurs in flattened or angular grains in beds
of gravel or sand, which resemble gold placers, and have
been formed in a similar manner by the erosion of lodes.
The richest and most extensive platinum placers are in or
near the Ural Mountains of Russia, in gravels 3 or 4 feet
thick, and buried below thicker layers of barren material.
Usually, but not invariably, the gravels are also auriferous, and other minerals occurring with the platinum are
zircon, spinel, corundum, magnetite, and osmiridium.

The deposits, besides quartz grains, contain fragments of basic magnesian volcanic and metamorphic rocks, such as serpentine, olivine rock, porphyries, etc. Platinum is recovered from the gold-bearing gravels of California and Oregon by dredging and hydraulicking. Alluvial deposits also occur in British Columbia, New South Wales and Colombia, the last named, after Russia, being by far the most important, and is being developed more and more. Much work has been done on the beach sands of the Pacific Coast, also on certain black sands of Western rivers, and prospectors have wasted much time thereby. It was definitely pointed out by the U. S. Bureau of Mines in 1919 that the beach sands of California and Oregon as a whole are not worth working for platinum, only in a few spots might a pocket be found.

Platinum occurs in place in serpentine in the Urals and elsewhere, and in other metamorphic rocks in various parts of the world, but the quantities so found are insignificant. It is also found as sperrylite (arsenide of platinum) associated with copper ores in the Sudbury district of Ontario. In the Rambler mine, Wyoming, where the ore produces palladium and one-third as much of platinum, basic dikes of diorite intrude granite-gneiss with some quartzite and massive granite. The dikes are much altered. A smelter, to reduce concentrate, was to be erected there in 1919. The mineral there is sperrylite. In the Boss mine, southern Nevada, the gold-platinum-palladium ore consists of fine-grained quartzose, which replaces the carboniferous dolomites, carrying a bismuth-bearing variety of the rare mineral, plumbojarosite (a lead-potash-iron-alum). In 1917, 443 tons of ore yielded 0.24 oz. of platinum and 0.9 oz. of palladium per ton. In

British Columbia, the best platinum was in dunite altered to serpentine, or in which the rock was rich in chromite.

Platinum in Rhodesia occurs in the Somabula diamond-bearing gravels. These are found almost on the main watershed of Southern Rhodesia, about 12 miles south-west of Gwelo. The pebbles also contain chromite.

Normally, Russia contributes 90% of the world's platinum, or 290,000 oz. in 1913. Colombia produced 27,-000 oz. in 1918, the United States a few hundred ounces, and Australia, Canada and others smaller quantities. The price of platinum has risen considerably in the past few years. In April, 1905, it was $20.50 per oz.; in February, 1906, it was $25, and in September, 1906, $34. In 1914 it was around $40, and during the period 1915 to 1918 was up to $110, being $130 late in 1919.

The name platinum is derived from *plata,* the Spanish word for silver, since it was regarded in South America at the time of its discovery — 1735 — as an impure ore of that metal.

Platinum, like gold, does not readily combine with other metals, and in nature the only compound known is an arsenide called *sperrylite.* Platinum may be distinguished by its great weight, by its gray color, its sectile nature, and by the fact that it will not dissolve in any simple acid, and with difficulty in aqua regia. It may be distinguished from lead by its action under the blowpipe flame, since lead melts immediately, leaving a yellowish coating, while platinum refuses to melt under the hottest flame, and leaves no coating whatever. When it exists in the alluvial soil it may be panned just as gold or other heavy metals, and even more easily because of its greater gravity.

A chemical test is as follows: Dissolve the grains of ore in 3 parts muriatic acid and 1 part nitric acid, preferably with gentle heat, add proto-chloride of tin solution [also called stannous chloride ($SnCl_2$)]; if platinum is present a dark brownish-red color will be produced, but no precipitate.

The metal may be obtained separate from gold admixed and in the presence of many other metals, by evaporating at a gentle heat the above solution of the ore in a porcelain dish to dryness with ammonium chloride (sal-ammoniac or muriate of ammonia), and the residue treated with dilute alcohol (one-fourth part water). The gold will remain in solution and the platinum be precipitated. The precipitate is ignited, when the platinum will be pure. Gold, if present, may be precipitated by adding a solution of ferrous sulphate [sulphate of iron (copperas crystals)] after evaporating off the alcohol.

Stannous chloride may readily be purchased at any dealer's in assayers' goods, but as it is easily prepared the best method is as follows: File a piece of tin into powder and heat nearly to boiling with strong hydrochloric acid in a porcelain dish or glass beaker, always keeping tin in the glass or dish, by adding tin if necessary. When no hydrogen gas is evolved that is, no bubbles arise, dilute with four times its bulk of pure water, slightly acidulated with hydrochloric (muriatic) acid, and filter. Keep the filtrate in a well-stoppered bottle in which some tin has been placed. If you have pure tin-foil, that form of tin may be used, for without the presence of metallic tin the stannous chloride ($SnCl_2$) is in danger of changing into stannic chloride ($SnCl_4$) with precipitation of a white substance (oxychloride of tin), which renders the reagent unfit for use.

Sperrylite. This is the only mineral known in which platinum occurs in combination with other elements. It is composed of platinum, 55.47% ; rhodium, 0.68% ; palladium, trace; antimony, 0.54% ; and arsenic, 43.23%. Color is tin-white; luster is bright; hardness is 7; and specific gravity is 10.6.

Platinum is more or less always associated with three metals of its group, namely:

Iridium, a steel-white, extremely hard metal, next in specific gravity to osmium, is supplied partly from its alloy with native platinum and partly from the iridosmium which occurs in platiniferous gravels. It is used for penpoints and in jewelry, and in metal-plating.

Osmium is the heaviest known metal. It comes from the same sources as iridium, and in the form of iridosmium is used for pointing tools and pens.

Palladium is a brilliant silver-white metal. It also occurs with platinum, but on account of its high price is but little used. A new compound called ' palau ' is now used sometimes as a substitute for platinum. It consists of 80% gold and 20% palladium.

CHAPTER VII·

COPPER

Copper occurs both native and as an element of various minerals. It is one of the most important metals, the world using over 1,000,000 tons yearly.

Native copper occurs as grains, filiform shape, plates, masses, and even in octahedral crystals. Color, copper-red, but often tarnished; ductile and malleable. Streak, shining; hardness, 2.5 to 3. Specific gravity, 8.5 to 8.9, according to purity. It frequently carries silver. Before the blowpipe, in the borax bead, oxidizing flame, small quantities are green while hot and blue when cold; large quantities are dark green while hot and greenish-blue when cold. In the reducing flame, small amounts are green while hot and greenish-blue when cold; large amounts are opaque, sealing-wax red, color strongest by candle-light. The salt of phosphorus bead gives the same result.

Native copper dissolves readily in nitric acid, and if ammonia be added the solution becomes green, or greenish-blue if it is in excess.

In the absence of any chemicals or a blowpipe, a mineral, when containing native copper, or when only a compound containing copper, may be tested by heating it either in the mass or, better, in powder, and when hot, dropping it into some salty grease and then putting it in a

flame or upon burning charcoal. The characteristic green
color will appear in the flame.

If the mineral contains copper in considerable quantity, and is dissolved in nitric acid, the copper will be deposited immediately upon a strip of polished iron or upon
the end of a knife-blade, if either be dipped into the solution.

There are many copper minerals that possess economic
value. These include sulphides, antimonides, arsenides,
oxides, chlorides, bromides, iodides, carbonates, sulphates,
phosphates, silicates, arsenates, simple and compound,
hydrous and anhydrous.

Several of the more important ores of copper are mentioned below and also some copper minerals, which, to
the prospector, will be suggestive that the more important
ores may be nearby.

Cuprite, Red Copper Ore, or Ruby Copper. Occurs
massive, granular, and earthy. Streak, shades of brownish-red, shining. Brittle. Color — deep crimson, cherry-
red. Luster, adamantine or sub-metallic; or again it may
be dull and earthy. It is sometimes weathered to an iron-
gray on the surface. Hardness — 3.5 to 4; specific gravity — 6. Composed of copper, 88.78%; and oxygen,
11.22%.

Before the blowpipe, on charcoal, it yields a globule of
metallic copper; with borax bead gives the reactions of
copper. Dissolves in hydrochloric acid, giving a brown solution, which, when diluted with water, deposits white insoluble cupric chloride. In nitric acid it forms a blue solution. Sulphuric acid decomposes cuprite into cupric
oxide (CuO) and metallic copper, the former passing into
solution as cupric sulphate, while the latter is undissolved.

Cuprite occurs in most copper districts as a secondary mineral in the oxidized zones. It frequently shades into crystals of native copper. Cuprite is an important source of copper. The massive variety is known as ' tile ore '; ' brick ore ' is a mixture of copper and limonite. The fibrous variety is known as ' plush copper ore.'

Chalcocite, Copper Glance, or Vitreous Copper. Massive; slightly sectile. Color — bluish-lead gray, brownish, brilliant when fresh, black and dull, on exposure to sunlight tarnishing to blue or iridescent. Streak — blackish-gray, sometimes shining. Hardness, 2.5 to 3; specific gravity, 5.5 to 5.8. Composed of copper, 78.8% ; sulphur, 21.2; and sometimes a little iron.

Before the blowpipe it gives off an odor of sulphur. When heated on charcoal, a malleable globule of metallic copper remains, tarnished black, but flattens under the hammer. With borax bead it gives the reactions for copper. Dissolves in nitric acid, forming a blue solution. These tests distinguish it from sulphide of silver. This mineral occurs in all copper districts, and yields half the world's supply of the metal. At Ajo, Miami, and Ray, Arizona ; Chuquicamata in Chile ; Ely in Nevada ; Santa Rita and Tyrone, New Mexico ; and Bingham, Utah, are the greatest copper deposits in the world. They have all been developed within the past dozen years, and contribute most of the world's copper. They are known as the ' porphyries,' and are low-grade ($1\frac{1}{4}$ to 2% copper) disseminated deposits of chalcocite, the result of secondary enrichment of lean sulphides scattered through highly altered porphyritic or schistose rocks. Weathering, erosion, and oxidation have combined to concentrate the copper into definite horizons. None of them are deep-seated, the Miami being perhaps the deepest — 900 feet. All are covered with overburden, up to 200 feet, and some are worked by open-cut methods and others by underground systems. The capping at Bingham, Ely, and Santa Rita is rich enough to be treated by a leaching process.

Tetrahedrite, or Gray Copper Ore. This is a sulph-antimonite of copper containing 52.1% of the metal. It is brittle; steel-gray or iron-black, sometimes brownish; with a streak between steel-gray and iron-black, sometimes brownish; hardness — 3.5 to 4.5; and specific gravity — 4.4 to 5.1.

Before the blowpipe on charcoal it fuses, gives an incrustation of antimonious, and sometimes arsenious acid, oxide of zinc, and oxide of lead. Arsenic may be detected by its odor on heating the incrustation in reducing flame, or fusing with soda. Oxide of zinc gives a green color when heated with nitrate of cobalt solution. The iron and copper in the residue are identified either by fluxes (on platinum) or by reduction with soda. Silver is determined by cupellation.

Tetrahedrite is soluble in nitric acid, arsenious and antimonious acids separating. The solution becomes blue from copper on adding ammonia in excess, and cloudy with hydrochloric acid when silver is present.

Tetrahedrite occurs with copper pyrite, galena, and blende. It is not regarded generally as a commercial ore of copper, unless it carries much silver, which is frequently the case.

Chalcopyrite, or Copper Pyrite. This is the primary ore of copper, and is second only to chalcocite as a source of the red metal. It is massive. Its color is brass-yellow, when fresh, gold-yellow when tarnished. Luster —

sub-metallic; brittle, slightly sectile; streak — greenish-black, slightly shining. Hardness, 3.5 to 4; specific gravity, 4.15. Composed of copper, 34.6%; sulphur, 34.9%; and iron, 30.5%. Before the blowpipe it fuses with intumescence and scintillation to a rough magnetic globule. When powdered and roasted at a low heat, it is converted into a fritted mass, giving reactions of copper and iron with fluxes. With soda on charcoal, gives a globule of metallic iron and copper. It is sometimes mistaken for gold, or iron, or tin pyrite; but it is brittle, while gold is not; it will not strike fire, as does iron pyrite; and it may be distinguished from tin pyrite by the film that the latter leaves on the charcoal. Chalcopyrite is soluble excepting the sulphur, in nitric acid, and on being heated yields part of its sulphur. Exposure to moisture and heat result in the copper and iron being converted into sulphates. Chalcopyrite alters to 10 other copper minerals, including malachite, bornite, brochantite, and chrysocolla.

Chrysocolla, or Copper Silicate. Accompanies other copper ores, occurring especially in the upper part of veins with carbonates. It is a bright green or bluish-green mineral, carrying 36% copper and considerable silica. In some districts it is an important mineral. Its hardness is 2 to 4, and specific gravity 2 to 2.3. Its luster is vitreous to earthy, and color is green to bluish-green. Its powder (streak) is white. When dissolved in nitric acid a residue is left; this distinguishes it from malachite, which is wholly soluble. Chrysocolla is decomposed by acids without gelatinization. Before the blowpipe with soda it gives a bead of copper.

Melaconite, or Black Oxide of Copper is usually

found on the surface. It soils the fingers when pulverulent. Its color and streak are black; luster is sub-metallic; hardness 3 to 4; and specific gravity 5.8. It occurs in masses of a dark, earthy appearance, sometimes in minute, shining particles. If the dusty powder be rubbed between the fingers and dropped on a flame, the latter will be colored green. It is soluble in ammonia, the solution being azure blue.

Malachite, or Green Carbonate of Copper, has a fibrous structure, nearly opaque, and is of an emerald-green color. It contains 57.5% copper. Streak — paler green than the color; hardness — 3.5 to 4; and specific gravity — 3.9 to 4. Commonly found near the surface of veins containing copper. Malachite is sometimes massive, granular, and disseminated as stains, this last feature often misleading prospectors into thinking that they have found a great deposit. When massive and well marked, malachite is used in the arts.

Before the blowpipe it becomes blackish. With borax it yields the usual blue-green bead, and on charcoal is reduced to metallic copper. It dissolves completely in nitric acid, and thus may be distinguished from silicate of copper, which has nearly the same color and will not dissolve.

Azurite, or Blue Carbonate of Copper, is an important source of copper, and is used for ornamental purposes. It is of a deep cobalt-blue color, sometimes transparent and brittle. Streak — bluish; hardness — 3.5 to 4; and specific gravity — 3.5 to 4. Can be scratched with a knife; and blackens when heated. On charcoal it is reduced to a globule of pure copper. With the borax bead it gives the indications of copper.

It is soluble in nitric acid with effervescence, forming a blue solution.

Azurite is often found in the oxidized zone of copper deposits, nearly always with malachite, but is less common than the latter.

Bornite, Variegated, Horseflesh, of Peacock Copper. This is another important source of copper. The Kennecott mine in Alaska and North Lyell in Tasmania have been wonderful instances of producers of this ore. It is a copper and iron sulphide, usually massive, of a copper-red to a bluish-brown color, with a blackish to lead-gray streak. Hardness, is 3; and specific gravity, 4.9 to 5.4. It contains 55.5% copper, 28.1% sulphur, and 16.4% iron. Before the blowpipe it gives a bead of copper.

Brochantite. This is a basic copper sulphate containing 56.2% copper, and is one of the principal minerals at Chuquicamata, Chile, where 11,000 tons of 2.1% ore is mined daily. It is also mined at Clifton, Arizona. Its hardness is 3.5 to 4; gravity, 3.9; luster, vitreous; color, emerald green; and streak, paler green.

Geology of Copper. Ores of this metal are found in rocks of almost every age. An outcrop is the surface expression of an orebody, so a correct understanding of the significance of surficial signs is of much importance in the search for mines, according to F. H. Probert of the University of California. Those interested in the study of outcrops should secure Bulletins 529 and 625 of the U. S. Geological Survey. Most of the copper deposits are genetically related, primarily, to either an acid-porphyry or monzonite, or to a more basic diabase. The original deposits formed directly from molten magmas are seldom of commercial value. Nearly all of the workable orebodies are the result of concentration and enrichment brought about by the agency of meteoric water, often repeated. There are several types of copper deposits — fissure-veins, contact metamorphic deposits, the so-called ' porphyry coppers ' or low-grade masses of disseminated mineral, magmatic deposits, replacements, and others. Each has its peculiar characteristics of surface expression, of metal and rock associations. The nature of the enclosing rock frequently influences the outcrop, so requires study. Too much significance should not be attached to attractive coloring, as a little copper may stain a wide area, and a little iron a whole mountain. Many worthless veins show large gossans. The porphyry coppers — such as Ajo, Chile, Chino, Ely, Inspiration, Miami, Ray, and Utah — today contribute the bulk of the world's copper supply. They are only the development of the past dozen years. These deposits are the result of secondary enrichment of lean sulphide scattered throughout highly-altered porphyritic or schistose rocks. The great Michigan lodes, which carry native copper, are not characterized by conspicuous outcrops. Altitude and climatic conditions affect copper deposits greatly.

The surficial leaching of the outcrop of a vein containing gold, copper pyrite and other copper minerals may so concentrate traces of gold as to make the outcrop valuable for that metal, while the leaching of the copper leaves a gossan, or sintery mass of ironstone, enriched as to gold but impoverished as to copper. This is exemplified by early mining at Butte, Montana, and Mt. Lyell, Tasmania.

The following are brief descriptions of the geologic occurrence of certain copper deposits throughout the

world.   There are many other small though interesting
ones, especially when associated with other minerals, but
these should suffice:

The Copper Country of Michigan produces up to 300,-
000,000 lb. of copper per year, and is the third district
in importance in North America.   Here the copper all
occurs native, from masses weighing many tons to the
finest particles.   The principal rocks are the upper, middle, and lower Keweenawan series, consisting of (1) conglomerates, sandstones, shales, and marls; (2) basalt
flows with inter-stratified sandstones, conglomerates, and
shales; and (3) conglomerates, sandstones, and shales.
The deposits in the second and lower part of the third are
the most important.   They have been opened to a depth
of over a mile, and are remarkably persistent.   Silver is
associated with the copper, but in small quantity, as in
1917, only 684,000 oz. were recovered from 268,500,000
lb. of copper.

Another deposit of native copper is that of the Corocoro
United mines in central Bolivia.   There the copper is
found native in several beds of sandstone inter-stratified
with reddish and brownish shale.   The ore-bearing beds
occur in 2 series, separated by a strong fault; the beds to
the east of this fault are known as ramos and dip east, the
copper-bearing strata being succeeded by a considerable
thickness of red slates capped by green slates and limestone overlaid in turn by red clay and brown sandstone.
The beds west of the fault are known as vetas, or veins,
and are much harder than the copper-bearing ramos.
The mineralization is not uniform and the native copper
occurs in particles scarcely visible to the naked eye, as well
as in grains, plates, threads and sometimes flat masses

weighing several hundred pounds. The main fault, known as Dorado vein, is also mineralized and is now the main producer of the company, yielding sulphides and arsenides of copper. It is supposed that this fault was the feeding fissure whose solutions spread into the sandstone and was precipitated by organic matter.

One of the new and important mines of North America is the Engels in California. There the orebody consists of large lenses in porphyry, alongside of diorite dikes, and well adapted for development by tunnel. The ore is principally bornite, with some chalcocite and chalcopyrite, and averages 2¾% copper. A capping of carbonate ore overlies a portion of the orebody, with sulphide ore at shallow depth.

At Rico, Colorado, ore occurs in fissure veins as replacements of limestone. In one mine it carries 4% copper, 10% lead, 10% zinc, and 10 oz. silver, the ore being chalcopyrite associated with pyrite, sphalerite, and galena in a fluoritic gangue.

Ore in the National mine, Idaho, occurs in a fault vein in thickly-bedded quartzite, with talcose slips bordering the ore.

Black and red oxides and chalcopyrite, carrying silver and gold, are the copper minerals in veins in andesite at the Bonney mine, New Mexico.

At Gold Hill, Utah, the predominant rocks are granite and limestone, and the deposits are replacements in limestone. The ore averages 4½% copper and nearly 1 oz. of silver for each per cent. of copper.

At the Pilares mine at Nacozari, Sonora, Mexico, one of the world's great mines, the principal rocks are latite, andesite tuffs, and breccias. These are cut by intrusions of monzonite, diorite, quartz-porphyry, and a diabase dike. The copper minerals are chalcopyrite, chalcocite, and bornite. The surficial indications of the orebody are uncommon. In the surrounding gray latite are reddish-brown knobs rising above a red-stained area. They contain specularite (an iron mineral), quartz, sphalerite (zinc), chalcopyrite, and cuprite. In addition to stains of limonite and manganese oxide they are smeared with malachite. The knobs represent excellent indicators of ore-bearing solutions, and looked promising to early prospectors.

The copper deposits of the Yerington district, Nevada, which yielded 80,000,000 lb. of metal from 1912 to 1918, consist of chalcopyrite and pyrite inter-grown with pyroxene, garnet, and epidote. They are of the contact-metamorphic type, and their geologic features are of much scientific interest. The investigation of the origin of the ore deposits has led to certain practical recommendations as to the most systematic and effective way in which to explore for undiscovered ore.

The Butte district is the largest base-metal producing center in the world, yielding copper, gold, lead, manganese, silver, and zinc. On the surface large quartz veins form a number of abrupt ridges and narrow crests rising above the general slope, while smaller veins form broken-down walls or reefs. The outcrops are composed in part of true vein-quartz, and in part of an altered granite indurated by secondary silica in minute veinlets and disseminated through the substance of the rocks, according to W. H. Weed. They are generally black with manganese oxides or rusty with iron. Such veins, though long and of great width, have not been great mineral producers. Many of the lodes now worked extend upward to the outcrop. At the East Butte mine there are no outcrops at all. The productive deposits are practically confined to the quartz-monzonite or Butte granite.

The great Bonanza and Jumbo mines of the Kennecott Copper Corporation in Alaska are in magnesian (dolomite) and non-magnesian limestone, above the greenstone contact. The ore mineral is dominantly chalcocite.

A feature of the two big mines at Jerome, Arizona, is that the surface carries large jasper outcrops, and assay well in copper, silver, and gold. The United Verde ground shows granular porphyritic igneous rocks, slate and schist, the ore being connected with intrusions of acidic porphyry in dioritic rocks, which are sheared and schistose. The overlying limestone has no connection with the ore, that is, as a source of it.

The great Wallaroo and Moonta lodes of South Australia traverse schistose rocks of sedimentary origin and are in brittle feldspar-porphyry, respectively. The important minerals are chalcopyrite, pyrite, chalcocite, and a little gold and silver.

Mt. Lyell, Tasmania, rocks are mica-schist, conglomerate, quartzite, and sandstone. The outcrop was a huge and rich gossan, and the principal mineral is a low-grade pyrite. In the North Lyell, the ore is a dissemination in quartzose-schist, bornite being the chief ore.

Mt. Morgan, Queensland, is an instance of a gold deposit turning out to be the silicious gossan of a great copper deposit, unlike anything yet discovered, not having any great length like other copper lodes, and has paid $45,000,000 in dividends.

In the Superior district of Arizona, at the Magma

Chief mine, one vein has a wide manganese outcrop, and all the known mineralized faults show the mineral and iron-stained outcrops of silicious material.

In the Shasta County copper belt of California, where the ores are almost all sulphides, the upper portions of the orebodies, at surface and at shallow depth, consist largely of impure limonite. Some of the deposits are covered by an 'iron hat,' and the graduations at depth are the gossan or oxidized section, underlying enriched sulphides, and unaltered sulphides beneath. The gossan contains some ($2 per ton) gold and silver, and is useful as a flux. Eventually it might be worked as an iron ore.

The Penn mine, Calaveras county, California, is a good little copper producer. Ore there lies between amphibo-lite-schist and talcose-shale. The chalcopyrite is accompanied by pyrite and sphalerite, with fair value in gold and silver. The gangue ranges from talcose-schist, through clay, to quartz.

In Spain, the Cordoba Copper Co.'s veins are in mica-schist, diorite, and quartzite, carrying chalcopyrite and pyrite, with gangue of calcite, quartz, and country rock.

At Cerro de Pasco, Peru, the orebodies outcrop prominently as ridges. To a depth of 100 feet, gold and silver contents are high, followed by silver-copper ore, then copper-silver ore. Galena, sphalerite, and bismuth occur in fair quantities.

Veins of the Ashio mine in Japan traverse a Tertiary liparite (rhyolite), and carry the usual copper minerals, with quartz and calcite as gangue.

It is well to remember, especially when exploring a new country, that copper is frequently associated with rocks of a dark color, which are often green; but it must not be supposed that the color is imparted by copper, as it is generally due either to some other metal, such as iron, or to the presence of a green non-metallic mineral, such as chlorite.

A notable exception, however, is to be found in the disseminated deposits in the vicinity of Globe and Ray, Arizona. Here the rock carrying the mineral is chiefly the very light-colored Pinal schist.

Serpentine and hornblende rocks are often associated with copper ores, but green serpentines owe their color to iron, nickel, or chromium, and if copper is found disseminated through some of them, it is the exception and not the rule, unless in the immediate vicinity of ore deposits. On the contrary, iron and chromium are found in all serpentines, and nickel is of frequent occurrence.

All copper ores weigh more than quartz or limestone, and the comparative weights should be so well known by practice that there should be no hesitation in judging that the mineral in hand is more than 2.6, which is the specific gravity of quartz or limestone.

Next examine the mineral with a pocket lens for any evidence of copper, such as green or bluish spots, or brassy points or particles. If these are found, chip one off and use the blowpipe with borax bead or with soda or borax on charcoal. If the characteristic color appears, it is copper. Now proceed with other parts of the specimen. If there is a distinct odor of sulphur, it is probably a sulphide. Place a small chip upon a depression in the charcoal, cover with soda or borax, turn the inner flame upon it and reduce to a metallic globule; if it shows the color of copper and is malleable, it is that metal;

if it blackens, apply a magnetized knife-blade, and if attracted, the mineral contains iron, perhaps both this and
copper.

The next work is to examine the region to gather any
other specimens and evidences of true ores, before at-
tempting to know more of any particular sample. If the
surface specimens are numerous, it may be well to collect
a half dozen, and proceed to examine them for the available copper. This is the work of a chemist, and should
be submitted to him; but as the skilled prospector frequently wishes to be his own chemist, where work for the
desired object is not difficult nor complicated, the following simple process will allow of estimating the percentage
of copper in an ore without regard to other elements contained therein: The only chemicals needed are nitric
acid, ammonia, and sodium sulphide — the colorless crystallized hydrosulphide of soda of commerce is good
enough. All the apparatus needed is a glass flask or tall
glass beaker and a graduated glass tube called a burette.
This may be obtained at any chemical supply-house. The
burette is marked in cubic inches or cubic centimeters,
from 25 to 100. Dissolve some sodium sulphide in clear
rain-water — about a half ounce to a pint. Keep the solution in a glass-stoppered bottle. Obtain some pure copper (ordinary good copper wire will serve), weigh accurately and dissolve in nitric acid, add some water (twice
the amount of acid used, or a little more), then ammonia
until, when stirred with a long piece of glass rod, the
solution smells strongly of ammonia. The ammonia must
be in excess. Now fill the burette with sodium sulphide
to the 100-mark, and run the solution from the burette
into the copper solution until the blue coloration entirely

disappears. Note on the burette by its marks the exact amount of sodium sulphide used. That amount represents the weight of the amount of copper used.

Pulverize some average ore, weigh it, and treat it as was done with the copper — with nitric acid and ammonia — and proceed with the sodium sulphide. When the ore solution has become entirely colorless, note what amount of sodium sulphide solution has been used, and the exact amount of copper in the ore may be calculated by simple proportion. The presence of tin, zinc, lead, iron, cadmium, antimony, arsenic, or bismuth in the ore does not interfere with the operation; but silver does. Therefore, a small amount of ore must be dissolved in chemically pure nitric acid, and tested by putting into the solution a drop or two of hydrochloric acid, or solution of common table salt (sodium chloride). If any silver exists in the ore a milky cloudiness will appear, of a density greater or less in accordance with the amount of silver present; if no silver appears, then the test may proceed as already directed. If silver does appear, then the solution containing the weighed ore must first be treated with the salt solution or diluted hydrochloric acid, until all cloudiness or white precipitate entirely ceases. The solution of ore now contains no silver, and the test may go on.

This process will give fairly accurate results, provided the following precautions are observed:

(1) Heat the copper solution, after adding the ammonia, to boiling point or little below while adding the sodium sulphide. (2) Add a little ammonia to the ammoniacal solution to replace that lost by evaporation. (3) When the blue ammoniacal solution begins to lose its color, drop the sodium sulphide in cautiously, so as not

to exceed the amount necessary to precipitate exactly the copper and no more.

The sodium sulphide first produces a black precipitate of copper sulphide, but before that takes place the ammonia will give another precipitate, provided that the copper contains any lead or tin. If it contains zinc, that will be precipitated immediately following the black copper sulphide, but will be white. If it contains any cadmium, that will be precipitated at the very moment the de-coloration takes place, if adding of the sodium sulphide is continued. Cadmium is identified by a beautiful clear yellow precipitate. With care and skill each may be distinguished.

In simply determining the amount of copper, however, these precipitates need not be regarded; only pay attention to the point of de-coloration.

The sodium sulphide may need proving to see if it has lost any of its strength if kept for a long time. This may be done by a trial with a new solution holding a known amount of copper; or, exactly the same weight of crystals of sodium sulphide to the same amount of pure water may be used as before, and the old solution thrown away; or, by re-testing the sodium sulphide the same solution may be used for a long time, and if it has become weakened, make allowance for the additional sodium sulphide required. It should be kept in a cool place, away from the sun and light.

CHAPTER VIII

LEAD, ZINC, AND ANTIMONY

These three metals are nearly always found together, so will be considered in that order.

LEAD

This metal rarely occurs native, and then only in small quantities. It has been found in this form in Sweden. Specimens found in America have been doubted. Hardness, 1.5; specific gravity, 11.3 to 11.4. The most important ore of lead is the sulphide called

Galena. When chemically pure it contains 86.6% lead and 13.4% sulphur. Specific gravity, 7.2 to 7.5, according to admixtures. Color and streak, lead-gray. Luster, shining metallic; the exposed surface may be dull from tarnish, but the fracture is brilliant. Easily recognized by the characteristic cubical cleavage, which is easily obtained, or by the granular structure when massive. Frequently associated with other metallic sulphides, such as pyrite, chalcopyrite, arseno-pyrite, and blende. It occurs in veins, the gangue of which is either quartz, calcite, barite, or fluorspar, in granite and nearly all varieties of rock; but the larger deposits are usually found either in veins or in pockets, often of great size, in limestone strata. It is also found in gold-bearing lodes, in which cases the ore is generally of good value, the galena being considered an indicator of such.

Galena nearly always contains silver, hence it should be tested according to the method given on p. 158. What might be termed an exception to this rule are the lead-zinc deposits of Arkansas, Kansas, Illinois, Missouri, Oklahoma, and Wisconsin. It has been stated that galena with small crystalline facets, like coarse lump sugar, is rich in silver, while that with large cleavages is poor; but this characteristic at best is only local, as some galena with large cleavages yields as much as 1,500 oz. of silver per ton, while other fine-grained ore contains only 50 oz. per ton, or even less.

Carbonate of Lead, White Lead Ore, or Cerussite. If perfectly pure, its composition is, lead, 83.6%; and carbon dioxide, 16.4%. Hardness, 3 to 3.5; specific gravity, 6.4 to 6.5. Color, white to gray, or even black if it has been much weathered. Streak, colorless; luster, glassy or adamantine; when pure is translucent, or even transparent. It is very brittle. If it contains copper it is usually tinged blue or green. It has a glassy or vitreous appearance, is easily melted before the blowpipe, and a lead bead or globule is readily obtained.

By placing a little bone-ash in a hollow in a stick of charcoal, and turning the oxidizing flame on the lead, a little skilful blowing will cause the lead to be absorbed and drawn off, leaving a bright silver globule, provided the lead contains silver. This is called blowpipe cupelling.

Carbonate of lead occurs in compact, earthy or fibrous masses, and is often found near the surface of a galena lode, being a secondary mineral derived from the alteration of galena. It is not a commercial ore of lead. At Leadville, Colorado, the carbonate has been found carrying 20 to 60% lead and 600 to 20,000 grams (20 to 670 oz.) of silver per ton, the silver being present as a chloride.

Sulphate of Lead or Anglesite. This often accompanies the carbonate. It somewhat resembles the latter, although it is of slightly less hardness — 2.75 to 3; with specific gravity, 6.12 to 6.39. It is often in rhombic crystals. Luster is adamantine or glassy; streak, white, gray, or black; and fracture, conchoidal. It may be distinguished from the carbonate by the fact that it does not effervesce in an acid, as the latter always will. It is composed of lead oxide, 73.6%; and sulphur trioxide, 26.4%, in pure specimens. This mineral is a common oxidation product of galena, but is only found in small quantities as a rule. The Cerro Gordo district of California is a good example of deposits of lead sulphate. Anglesite contains 68.3% of lead.

Phosphate of Lead or Pyromorphite. This is another oxidation product of galena. Composition, when pure, 89.7% phosphate and 10.3% chromate of lead, with arsenate of lead up to 9%, phosphate of lime, 0.11%, and calcium fluoride. Hardness, 3.5 to 4. Specific gravity, 6.5 to 7.1. Color, greenish, sometimes bright grass-green, the hexagonal crystals having a greasy luster, also yellowish, brownish, and sometimes dull violet. Luster, resinous; generally translucent. Streak, white or yellowish. Contains 78% lead. Heated on charcoal before the blowpipe a globule is formed, which takes on a crystalline appearance on cooling, leaving a yellow oxide of lead on the charcoal. With carbonate of soda in the reducing flame it yields a yellow globule. It is soluble in nitric acid.

Crocoite or Chromate of Lead. This is a yellow mineral containing oxide of lead, 68.9%, and chromic acid, 31.1%. Hardness, 2.5 to 3; specific gravity, 5.9 to 6.1 Color, various shades of bright hyacinth-red. Streak (powder), orange yellow. Luster, vitreous, translucent, and sectile. Decrepitates before the blowpipe. With soda on charcoal it yields a lead coating; with borax an emerald green bead.

Massicot or Lead Ocher. This mineral occurs massive, as a compact earth of a sulphur-yellow or reddish-yellow appearance. Hardness, 2; specific gravity, 8, and, when pure, 9.2. It is composed of oxygen, 7.17%, and lead, 92.83%. Before the blowpipe it fuses readily to a yellow glass, and on charcoal is easily reducible to metallic lead.

Lead-Antimony Ores. There are several compounds of lead with antimony, but they are never sufficiently plentiful to be considered as ores. One of these — jamesonite - contains small quantities of iron, copper, zinc, and bismuth. The mineral is a sulphide of lead and antimony. It occurs in gray fibrous masses or small prisms, and is found in Cornwall, England, associated with quartz and bournonite (a sulph-antimonate of lead and copper). Jamesonite is also found in Arizona, Nevada, and South Dakota. Another of these compounds — zinkenite — resembles stibnite and bournonite, and occurs in an antimony mine in the Harz, Germany.

Geology of lead. The ores are found in veins, lodes, flats and beds, and in pockets (Fig. 52). Galena occurs in limestone, a yellowish-gray, hard, compact crystalline rock. The lowest horizon of lead ore in workable quantities lies above that of copper.

Lead is nearly always associated with zinc, notable examples being the Tri-State region of Kansas, Missouri, and Oklahoma, the Cœur d'Alene region of Idaho, Broken Hill in Australia, and Bawdwin in Burma. Exceptions to the rule, where clean lead ore is mined, are in Southeastern Missouri, Bunker Hill & Sullivan, and others in Idaho.

In the most important lead region in the world —South-

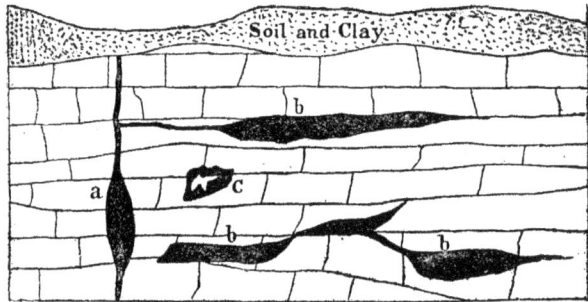

Fig. 52.

Section of Galena limestone showing how lead occurs in lodes, *a,* flats, *b, b, b,* and pockets, *c,* from mere threads to several feet in thickness.

eastern Missouri — the deposits are in dolomite, to a depth of 700 ft. The ore carries from 4 to 6% lead, but no silver. In the Bunker Hill & Sullivan mine, most of the ore is a replacement of a quartzite, the shoots being definitely related to a persistent figure. The ore carries silver. In the Utah Consolidated the deposits are replacements of limestone beds, with quartzite above and below. The galena is associated with copper minerals. In Wisconsin, the ore deposits occur in cracks or crevices in the limestone, are in the nature of veins; also disseminated in small particles throughout the limestone. The formations in which the orebodies are found lie under the following: residual soil, alluvium, and clay, 7 ft.; dolomite, 100 ft.; shale, 160 ft.; Galena limestone, 230 ft.; and Platteville limestone, 55 ft. thick, contain the productive deposits.

Water has much influence on changes in the physical character of veins and the circulation of water in the veins of British mines is affected by the inclination of the strata in the direction of the vein. The richest deposits are found in that portion of strata that is the most elevated; for instance, on the side of a strong cross-vein. The circulation of water is dependent upon an outlet at a lower level. In the case of lead mines, it is stated that in consequence of conditions connected with the descent of water, the richest deposits are generally found at no great distance from the outcrop of the containing rock. Veins that occur on the side of a mountain in a direction nearly parallel with the valleys contain more extensive deposits of lead than those that cross the valleys at right angles. The prospector should keep this suggestion in mind. Lead ores are found in fissures where they seem to have been deposited by waters that have dissolved them out from neighboring beds.

Prospecting in the Tri-State region used to be done by sinking shallow shafts or test pits. This method became too expensive for the deeper deposits, so churn-drills are used now. In the Wisconsin fields, the most favorable points for prospecting are the areas that produced lead ore near the surface, although below water-level there will be less lead, and probably more zinc.

Lead has been the principal product of the Tintic district, Utah, but the output of gold and silver, with some copper and zinc, is also important. A large part of the surface is covered by a thick mantle of disintegrated rock, making prospecting difficult, as there are few outcrops and indications of mineralization. The principal veins are in Paleozoic limestone and dolomite, but monzonite, porphyry, and rhyolite also contain ore-bearing veins.

In the Simon district of Nevada, attracting much attention during 1919, outcrops of gold, copper, and lead are frequent. Oxidation and the action of surface water have altered and leached the surface of the veins to such an extent that little of the metallic content remains, necessitating sinking to the unleached part — usually at 250 to 400 feet depth. Contact veins are frequent, and silver-lead veins largely follow the contacts between limestone and the intrusive igneous rocks — granites, granodiorites, and diabase. Many cross fissures radiate from contacts, containing good ore. The mineralized area extends far into the limestone.

ZINC

This metal is never found free in nature, but chiefly occurs in combination with carbon dioxide or sulphur. The chief ores are:

Smithsonite or Zinc Carbonate. Composition — zinc, 52.06%; oxygen, 13.10%; carbon dioxide, 35.20%. These proportions vary on account of the presence of protoxide of iron, manganese, and magnesia. Color — when pure, is nearly white, through various shades of yellow and gray to brown. Hardness, 5; specific gravity,

4.3 to 4.4. Streak — uncolored or white. Luster — vitreous, pearly, sub-transparent to translucent.

In Wisconsin the miners call it ' dry bone.' It is found in porous masses in limestone, forms thin coats on the rock, and often replaces calcite. Smithsonite is a secondary mineral usually associated with lead carbonate and silicate of zinc. It is not used as a source of spelter but for making zinc white, used in paint.

It is easily detected by the blowpipe, as it gives a green color when heated, after being moistened with half a drop of nitrate of cobalt solution. On charcoal, with soda, it coats the charcoal with a white film, which is yellow when hot and white on cooling; but if moistened with the cobalt solution and heated in the oxidizing flame it turns green. With muriatic acid it effervesces and dissolves. In mass it is translucent and brittle.

Calamine. This is a silicate of zinc, composed of zinc oxide, 67.5% ; silica, 25% ; and water, 7.5%. Hardness — 4.5 to 5, the latter when crystallized (Dana) ; specific gravity, 3.4 to 3.5. Color — when pure, pearly white ; but owing to the presence of iron oxide and other minerals, is generally brownish, sometimes green. Streak —whitish. Luster — pearly or glassy. Acts before the blowpipe like smithsonite, but does not effervesce with acids, and gelatinizes ; it is soluble in a strong solution of potash. In physical characteristics zinc silicate somewhat resembles zinc carbonate, and is often confounded with it. An anhydrous variety of this ore is willemite, which is plentiful in the New Jersey Zinc Co.'s Franklin mine. Zinc silicate is usually found in veins or in beds or in irregular pockets in stratified calcareous rocks, in association with zinc blende, zinc carbonate, iron, and lead ores. The

Joplin district of Missouri and Northern Arkansas yield large quantities of calamine.

Zincite or red oxide of zinc. Its composition is zinc, 80%; and oxygen, 20% varied by the presence of 3 to 12 parts of peroxide of manganese, which gives the red color, as zinc oxide when pure is white. Hardness — 4 to 4.5; specific gravity — 5.4 to 5.7; color, red and yellowish-red; streak the same; luster — brilliant; translucent, brittle. Occurs in grains or masses. Is found chiefly in Sussex county, New Jersey.

Sulphide of zinc, sphalerite, or zinc blende. Throughout the Tri-State region and Wisconsin this important mineral is known as 'black jack' or 'jack.' Composition — zinc, 67.15%; sulphur, 32.85%, but varied by iron, and sometimes cadmium. Color varies from yellow to brown and almost black, having a waxy look. Streak — white to reddish-brown; luster — waxy. Hardness — 3.5 to 4; specific gravity — 3.9 to 4.2; brittle, translucent.

Geology of Zinc. Blende is the most abundant zinc mineral. It occurs in rocks of all ages, in veins, in contact deposits, or in irregular pockets in limestone, and is nearly always associated with lead. Other minerals found with it are copper, iron, silver, and gold. Gangue minerals are quartz, barite, fluorite, and calcite. The lead occurrences in Kansas, Missouri, Oklahoma, and Wisconsin have been described and will suffice for the zinc. Those are all blanket deposits. Instances of lodes rich in zinc are (1) the Consolidated Interstate-Callahan in Idaho, where the veins cut through slate, which are intruded by dikes of porphyry and diabase. (2) The Butte & Superior in Montana, where the ore occurs in parallel or branching shoots, and consists of blende, galena, pyrite, and chalcopyrite with quartz, silicified granite, rhodochrosite and rhodonite, as vein replacement of the granite country rock. The ore carries 17% zinc and 7 oz. of silver per ton. (3) The Sullivan mine in British Columbia, where the ore is a replacement deposit in a fine-grained argillaceous quartzite. (4) The Bawdwin mine in Burma, India, where the deposit is a replacement of a fine-grained dark-colored silicified rhyolite, the impregnated zone being over 100 feet wide in places. Walls are not well-defined. Ore is an intimate and complex mixture of sulphides averaging 19% zinc, 27% lead, and 24 oz. of silver, the last being argentite. And (5) the Barrier lode at Broken Hill, Australia, where the lead-silver-zinc ore, with a gangue of rhodonite, calcite, and quartz, is in schist.

Sphalerite is easily recognized if treated with hot hydrochloric acid, as it gives a smell of rotten eggs (sulphuretted hydrogen), and the same results can be obtained without heating if a small quantity of pure iron filings is added to the acid. With soda on charcoal before the blowpipe, blende gives up sulphur, which, with water on a silver coin, tarnishes or blackens it.

Its occurrence and that of lead are so nearly alike that what has been said of the latter will apply to the former.

In New Jersey a section of strata near Sparta, Sussex county, shows slaty rock with feldspathic dikes, then limestone adjoining the franklinite iron ore with zinc 20 to 30 feeet wide, then the red oxide of zinc 3 to 9 feet wide, then crystalline limestone, and next feldspathic rock.

Franklinite is an ore somewhat resembling magnetite in color, hardness, and specific gravity, but it contains manganese and zinc, and as an ore, is peculiar to New Jersey. Its color is dark-black, and streak, dark-brown. Its action on the magnet is feebler than in the case of magnetite. The iron is said to be of the composition of peroxide, or Fe_2O_3, but it is probably in part protoxide, and this is the cause of its feeble effect on the magnet. It is easily affected under the blowpipe. Alone, it is infusible, but with borax in the oxidizing flame it colors the borax bead with the amethystine color of manganese, and in the reducing flame it shows the bottle-green of iron. On charcoal with soda it gives the bluish-white manganate, and also the coating of zinc especially if the soda is mixed with borax. In fine powder it is soluble in hydrochloric acid.

ANTIMONY

The most important deposits of this useful metal are in China, France, Italy, and Mexico. American reserves are small. Chinese metal controls the world market, and American operation must be cheap to compete with it. Antimony rarely occurs free; some has been found in California and some in Nova Scotia. The common mineral is stibnite — the sulphide. This contains 71.4% metal. It is lead-gray in color, has a dark-gray streak (which is pale yellowish-brown when rubbed), metallic luster, is often found as long (up to several inches) prismatic (acicular or needle) crystals, is 2 on the scale of hardness, and has a specific gravity of 4.52 to 4.62. Stibnite is soluble in hydrochloric acid, giving a slight crystalline precipitate of lead chloride if lead be present. Before the blowpipe, on charcoal, it fuses, spreads out, gives sulphurous and antimonial fumes, coats the charcoal with white oxide of antimony, which, when treated in reducing flame, tinges it greenish-blue.

Stibnite is widely distributed in Alaska, where it is an accessory mineral in many orebodies. Where it is the principal metal, gold generally accompanies it, also arseno-pyrite, in quartz veins in schist. Stibnite also occurs there in limestone, bounded by schist.

In New South Wales, Rhodesia and Western Australia, antimonial-gold ores are worked, and are generally accompanied by arsenical pyrite.

Good deposits have been mined in California, in Inyo and Kern counties, where it is found as veins in granite and metamorphic gneisses and schists; also associated with cinnabar galena, sphalerite, and chalcopyrite.

On Pine creek, Cœur d'Alene region, Idaho, a company is concentrating stibnite; while Nevada has several interesting occurrences, one not far from Austin, Lander county, being in quartz.

Some remarkable deposits occur in Iron county, Utah, as masses of radiating needles, which follow the stratification planes of sandstone and fill the interstices of a conglomerate. Stibnite is found in Sevier county, Arkansas, filling veins, with a quartz gangue, in sandstone.

Among European centers of production, the Bohemian mines are in granite and mica-schist; the Hungarian in granite — sometimes auriferous; the Styrian in dolomite, and the Turkish also in granite. In New Brunswick, antimony is mined in a quartz and calcite gangue in clay-slates and sandstones of Cambro-Silurian age.

Jamesonite is a lead-antimony ore containing 58.8% and 29.5% of these metals, respectively, the remainder being sulphur. It is often found in silver-lead deposits. Its hardness is 2 to 3 and specific gravity 5.5 to 6. At the Cortez Associated mine at Zimapan, Mexico, is some of this ore, a replacement in limestone cut by a porphyry dike. The average content is 8% lead and 4% antimony.

CHAPTER IX

IRON

This is one of the most abundant elements of the earth's crust, its distribution being materially aided by the fact of its forming two oxides of different chemical quantivalence. Native or metallic iron is rarely found; then almost entirely as meteoric iron.

Native or Meteoric Iron. This is found in small grains in the platinum bearing sand of Russia and in larger coherent masses in rocks in Canada. However, of greater interest is the occurrence of native iron in meteorites, which occasionally reach the earth. As a rule they do not exceed a few pounds in weight, and only a few weighing more than 220 lb. have been discovered. In 1870, large masses weighing from 10,000 to 50,000 lb. were discovered imbedded in basalt at Disko, Greenland. Peary brought back from his early polar trip a meteor weighing over 30 tons. It contained 92% iron and 8% nickel. Several meteoric masses containing small black diamonds have been found in Arizona. Meteorites always contain nickel and traces of cobalt, copper, and other metals; in fact, the presence of nickel is the criterion for the genuineness of a meteorite. Actual meteoric iron in which the presence of nickel could not be established has thus far been found only at Scriba and in Walker county, Alabama. In specimens of meteorites examined, the iron ranges from 67 to 94% and nickel from 6 to 24%.

Meteorites may be distinguished from metallic iron of other origin by peculiar markings which are produced when the surface is polished and treated with nitric acid. These markings are due to combinations of iron with nickel, and partly also with phosphorus, which are not decomposed, or only with difficulty by acids, and are imbedded in a crystalline state in the remaining mass of iron, the latter being quickly dissolved by the acid, while the former are not attacked for a long time.

The most important ores of iron are:

Magnetite or magnetic iron ore is found in octahedral or decahedral crystals; more commonly simply massive. Its streak and color are black, and luster is metallic. Composition, iron 72.4%; oxygen, 27.6%. Hardness, 5.5 to 6.5; specific gravity, 5 to 5.1. The ore is always easily attracted by the magnet, and sometimes is found capable of attracting iron, and is then called 'polaric' or 'loadstone.' In powder or small grains it is always attracted by a magnetized knife-blade.

The usual geologic position of magnetite is in the most highly metamorphic rocks, in which it probably represents the excess of iron oxide originally in the rock that was not taken up by silica. Occasionally it is found in layers, but in this country and elsewhere it forms whole mountains. Among other rocks in which it exists the following are the most important: crystalline limestone, chloritic, talcose, hornblendic, pyroxenic and hypersthenic schists; serpentine, diorite, and basalt. Specular iron is frequently associated with it. Magnetite comprises the bulk of the black sands of California and Oregon.

Magnetite is not acted upon by nitric acid, but hydro-chloric acid dissolves it when in fine powder and under long-continued heat.

Iron exists in magnetite as protoxide and peroxide, or FeO and Fe_2O_3, and upon this difference of oxides is based the action of important tests.

Red Hematite or Specular Ore is the peroxide of iron without the protoxide. This oxide is also called the sequi-oxide, or 1½ oxide, since iron combines with oxygen in the proportion of 1 to 1½ parts, or Fe_2O_3, and this is the highest proportion of oxygen the iron will combine with, and hence it is the peroxide, the peroxide and sesqui-oxide being the same in this case.

Specular ore is called red hematite from its color, which in some masses is so intensely red as to appear nearly black; but it may always be distinguished from magnetite by its red streak, and the blacker the ore the more decided is the red of its powder or streak. It is never magnetic. It has been found that in cases where specular ore showed any magnetic attraction, it was due to the fact that the ore contained some protoxide of iron.

Its hardness is 5.5 to 6.5; specific gravity 4.9 to 5.3; and composition 70% iron, 30% oxygen. The luster is metallic, but dull and earthy in some varieties.

Brown Iron Ore or Brown Hematite or Limonite. This has the same composition as red hematite, except that it has less iron and contains water in chemical combination, generally about 14%. Its color is brown, yellow, and coffee color; and streak is yellowish. Luster is dull or sub-metallic. Hardness, 5 to 5.5; specific gravity, 3.6 to 4. Limonite is compact, stalactitic, botryoidal, columnar, fibrous, and earthy—hence ' bog-iron ore ' from the last characteristic. When heated red-hot it loses its water and turns to a bright red, unless largely mixed with alumina and silica, when the red color is shaded. It is not magnetic unless heated with soda under the blowpipe, when it becomes metallic, as all iron ores do.

The amount of iron in a pure specimen is 59.8%, sometimes decreased by the presence of alumina, silica, magnesia, and other impurities, so that its average in many good mines is only about 35 to 36% iron. It is not so valuable an ore of iron as magnetite or red hematite. Limonite is the common alteration product of pyrite and most minerals containing iron, so is prevalent in most mineral districts, forming the gossan and brown capping of ore deposits.

Spathic iron ore or siderite is an iron carbonate, composed of iron protoxide 62% and carbon dioxide, or 48.2% pure iron. It is sometimes massive, with a crystalline structure. Its hardness is 3.5 to 4.5; gravity, 3.7 to 3.9; streak, white; color, gray or cream color, unless weathered, when it is brownish; luster, glassy or pearly; and translucent on edges, and thin plates or splinters. When in powder it effervesces with muriatic acid, especially when hot.

With the blowpipe in a closed test-tube siderite decrepitates, becomes blackened, and gives off carbonic acid gas. Before the blowpipe alone, held by forceps, it blackens and fuses. In the tube with muriatic acid it may be tested for carbon dioxide by lowering a lighted thread down into the tube, when the flame is instantly extinguished. The solution in the tube may be tested for iron by dropping a little ferricyanide of potassium solution into the hydrochloric acid solution, when it at once becomes a deep blue. This is a test of protoxide of iron, spathic ore being iron in the condition of protoxide only.

Black band ore is an argillaceous spathic ore of various dark colors, being largely combined with carbonaceous material. It is found extensively in Great Britain, near the apex of the coal measures. In America the black band ore is also associated with the coal measures, both in the anthracite and bituminous regions.

Geology of iron ores: This study may be divided into (1) the magnetites, which are always derived from the granite, gneiss, schist rocks, clay-slates, and, rarely, the metamorphic limestones; (2) the red hematites, which seem to be only an alteration derived from the magnetites, and belong to the same more ancient rocks as the latter; and (3) the brown hematites (limonites) which are derived from both the former, and are generally sedimentary.

Frequently in extensive magnetic regions, where the country back is mountainous, the brown ore has been formed in basins and knees and interlocked portions of the lower country, where ages of rains, storms and freshets have gradually transported and altered the magnetic ores of the upper regions and brought down these iron oxides to the low lands, where they have been arrested and settled down in beds of brown hematite. This seems to have been the history of all the hematitic limonite beds and deposits; they are on the lower levels when they were formed, although in later periods they may have been uplifted.

Iron ores are, therefore, to be found in three general geologic regions: (1) in the earliest rocks; (2) in the carboniferous; and (3) in the more recent or sedimentary rocks; and in accordance with their composition as magnetites and specular ores, as carbonaceous or black band and spathic ores, or as brown ores of the limonite order.

In ordinary cases, where the surface is covered with loose earth, it is common to search for magnetic iron ore with a magnetic needle or a miner's compass, and for preliminary examinations it is the principal instrument used, although the magnetometer is becoming more popular. In using the compass considerable practice is required. First examinations are made by passing over the ground with the compass in a. northwesterly and southeasterly direction, at intervals of a few rods, until indications of ore are found. Then the ground should be examined more carefully by crossing the line of attraction at intervals of a few feet, and marking the points upon which observations have been made, and recording the amount of attraction. Observations with the ordinary compass should be made, and the variation of the horizontal needle be noted. In this way material may soon be accumulated for staking out the line of attraction, or for constructing a map for study and reference. After sufficient exploration with the magnetic needle, it still remains to prove the value of the deposit by uncovering the ore, examining its quality, measuring the size of the orebody, and estimating the cost of mining and marketing it. Uncovering should first be done in trenches dug across the line of attraction, and carried down to the rock. When the ore is in this way proved to be of value, regular mining operations may begin. In places where there are offsets in the ore, or where it .has been subject to bends, folds, or other irregularities, so that the miner is at fault in what direction to proceed, exploration may be made with the diamond-drill.

The search for iron ores, is like that for petroleum, one of special knowledge, involving geologic study. In the Lake Superior region, where lie the greatest deposits in the world, the formations are determined by outcrops, · by magnetic lines, by drilling, open pits, and by the general geologic structure. They are generally associated with hills or ranges. The good ore, as a rule, is softer than the enclosing rock, and outcrops are uncommon; but the lean silicious parts of the formations withstood erosion and outcrop rather commonly, and serve as guides to the ores in the areas covered by drift. In the great Mesabi range, the ore is in ferruginous chert, which is bounded by slate, quartzite, granite, green schists, etc.

CHAPTER X

CHROMIUM, MANGANESE, MOLYBDENUM, TITAN-IUM, TUNGSTEN, URANIUM, AND VANADIUM

These are known as the alloy or ferro metals, from the fact that they are alloyed with iron in an electric furnace and used in varied proportions to harden or toughen steel. The ores from which they are extracted are termed 'non-metallic minerals.' The demand from 1914 to 1919 was enormous, and will always continue to be so, therefore the search for them should always be maintained.

CHROMITE

This is an oxide of chromium and iron, containing 68% of chromic oxide. It is generally black in color, has a metallic luster, and a grayish-brown streak. Hardness — 5.5; specific gravity — 4.32 to 4.57.

Geology of chromite. Excepting in one or two rare instances, chromite is found only in serpentine rocks. It may be found in pockets weighing a few hundred pounds up to several tons, or in fairly well-defined lodes. Generally the latter do not persist below 30 or 40 feet, although in Canada one deposit is several times this depth. Chromite may be traced on the surface by float ore, while frequently the serpentine must be examined closely to find deposits. As much as 1200 tons of float ore has been collected from one claim. What are known as dis-

seminated deposits are those that have particles of the
mineral scattered through it, averaging say 6% upwards.
Such ore must be concentrated before marketing. Ore
for shipment must contain at least 30% and should be
low in silica — under 10%. Beach sands along the Pa-

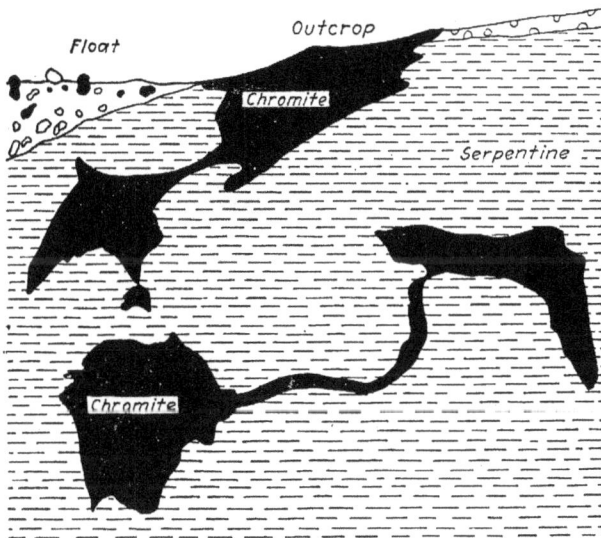

FIG. 53.— CHROMITE DEPOSITS.

cific Coast contain concentrations of chromite in spots.
California and Oregon are the principal producing States,
while New Caledonia and Rhodesia yield large quantities.

Chromite deposits are peculiar in that an outcrop, no
matter how small, may be the only surface exposure of
a large lens, or it may be all the ore there is. An open-

cut should first be started, then a shaft, following the ore-shoot.

In Cuba, all of the chrome exists in serpentinized basic rocks. The deposits are lenticular or tabular masses from 1 to 50 feet in thickness and a maximum of 200 feet in length. They resemble many of the occurrences in California and Oregon. The ore is not high grade, carrying from 26 to 43% chromic oxide, 10 to 13% iron, and 1 to 12% silica.

Field Determination of Chromite. Familiarize yourself with the appearance of typical chrome ore. Scratch a piece of the suspected ore with a knife, deep enough to penetrate any surface film of foreign material. If the streak is dark-brown, the material may be chrome. If the prospector will provide himself with a spirit lamp, two inches of platinum wire fused into the end of a short glass tube, and an ounce of borax, he can make a fair determination. The operation is as follows: Crush a small piece of the ore to a powder; bend a small loop at the end of the platinum wire; heat the wire in the flame, and dip it in the borax. Some borax will stick to the loop. Melt this in the flame, and continue dipping in the powdered borax and melting it until the loop has become filled with a bead of colorless borax-glass. Heat the bead to redness, and, while hot, place it against the crushed ore until a few particles adhere to the bead. Hold this in the flame until the particles are entirely dissolved in the borax, allow it to cool, and then the bead, if chromium is present, will have a bright green color. Having determined that chromium is present, send the sample to a reputable chemist and ask him to determine the amount of chromic oxide and silica. The ore-buyer must know the percentages of these two materials before he can go further. Magnetite, or magnetic iron ore, is frequently mistaken for chrome ore. It is dense black, highly magnetic, and of about the same weight as chromite. It is much harder, however, and can be scratched with a pocket-knife only with difficulty, and it does not show the brown streak. It may be distinguished also by crushing to the size of a pea. If it can be picked up by a magnet it is not chromite. Hematite, another ore of iron, is sometimes mistaken for chromite. It is usually softer, and has a reddish-brown streak, quite different from that of chromite. Hornblende-picrite, a rock that is dark-green to black, and quite heavy, is frequently found in chrome districts. Hornblende may be similar in appearance. The simple way to distinguish these rocks from chromite is to scratch them with a pocket-knife; the streak is grayish to greenish-white, according to Samuel H. Dolbear.

MANGANESE

The ores of manganese are divided into three general classes:

1. Manganese ores.
2. Manganiferous-iron ores.
3. Manganese-silver ores.

Wad is the name given to manganese oxide. It is found in earthy, compact masses of a dark-brown color, chiefly oxide of manganese and water (10 to 20%). It is easily recognized under the blowpipe, as it gives (in minute quantities) in the borax bead, a violet color in the oxidizing flame, but disappears when the reducing flame is turned upon it, and re-appears when the oxidizing flame is repeated.

Wad is found in beds varying from several inches to a foot or more in thickness. Its hardness is 0.5 to 2.5; specific gravity, 3 to 4; color, generally some shades of dark-brown; and streak umber-brown. It is generally soft and earthy, compact. Fracture is smooth, sometimes sub-conchoidal. Wad is used as a flux in iron smelting, and in a prepared state as a paint.

Pyrolusite. This is the peroxide or dioxide, with 63.2% of manganese and 36.8% oxygen. Its crystalline form is the rhombic prism and it generally occurs in the form of minute crystals grouped together and radiating from a common center. Pyrolusite is a common mineral, generally associated with other ores of manganese. It is frequently fibrous, and occurs with psilomelane. It has an iron-black or steel-gray color, a semi-metallic luster, and yields a dull-black streak. Specific gravity is 4.7 to 4.86; hardness, 2 to 2.5; infusible before the blowpipe, and acquires a red-brown color. On heating it generally yields some water and loses 12% of oxygen. With borax, soda, and microcosmic salt it shows manganese reaction. It dissolves in hydrochloric acid, when heated, with vigorous evolution of hydrogen.

Psilomelane occurs massive, stalactitic, occasionally shelly, seldom fibrous; color, iron-black to bluish-black; streak bluish-black to brownish-black and shining; fracture, conchoidal to smooth. Specific gravity, 3.7 to 4.7; hardness, 5.5 to 6. It is a hydrous oxide of manganese, and is nearly always associated with pyrolusite. Before the blowpipe it yields manganic oxide, giving off oxygen. It is soluble in hydrochloric acid, chlorine being evolved. The powdered ore colors sulphuric acid red. Psilomelane contains from 40 to 50% of manganese, and some baryta and potash. A solution in hydrochloric acid of the variety containing baryta gives a heavy white precipitate with sulphuric acid.

Rhodocrosite. This is a manganese carbonate and occurs in spherical and nodular aggregations of cauliform texture or in compact masses of granular texture. It is rose-red to raspberry-red in color, by weathering frequently brownish, with a glassy or mother-of-pearl luster. It cleaves like calcite, and has a vitreous luster. Rhodocrosite contains 61.4% of manganese protoxide and 38.6% of carbon dioxide, with part of manganese frequently replaced by calcium, magnesium, or iron. Specific gravity, 3.3 to 3.6; hardness, 3.5 to 4.5. Before the blowpipe it is infusible and becomes black. From similar minerals it is distinguished by its rose-color and the manganese reaction with soda and borax; and from silicate of manganese by its inferior hardness, its effervescence with acids and its non-fusibility. Butte, Montana, produces a large quantity of rhodochrosite.

Other manganese minerals are braunite — the silicooxide; manganite — the hydrous oxide; and rhodonite — the silicate. The last mentioned is rose-red, and is often present in copper and silver veins where manganese oxide is present.

Colorado produces two classes of manganese-bearing ores — a manganiferous iron ore used to some extent in the production of spiegeleisen, and a manganiferous-silver ore used as a flux in the smelting of silver-lead ores.

Manganese often occurs with silver. Such veins are prominent at Butte, Montana, and are worked for the manganese. They are north of the copper veins, but are closely associated with the zinc deposits. In Jalisco, Mexico, is a well-known manganese-silver orebody, but its. treatment is the subject of much experimentation.

The greater portion of manganese and manganiferous ores serves for the preparation of iron-manganese alloys — spiegeleisen and ferro-manganese, as well as other manganese alloys. Considerable quantities are used as coloring matter in the manufacture of pottery, as a scouring agent in making glass, and in dry electric batteries. Manganese is also used as a coloring matter and mordant in dyeing and calico printing, in the manufacture of oxygen, as a material in the manufacture of disinfectants.

Manganese, when heated in the air, gives a good brown paint, and when moderately heated, a black paint. Both paints may be produced directly from the residue from the manufacture of chlorine.

Most of the manganese deposits of the United States are relatively small, nothing comparable with those of Brazil, India, or the Caucasus of southern Russia. In the Batesville region of Arkansas the deposits are in shale and limestone (locally termed ' gray rock '), and owing to erosion and solution of these rocks most of the ore is in the residual clays of these formations. These clays are of red, yellow, and chocolate color, and contain some chert. The ores are irregularly distributed, ranging in size from fine particles to boulders over 20 tons. The minerals are the oxides, chiefly psilomelane and braunite. These ores are suitable for ferro-manganese manufacture, and contain 45 to 52% manganese, 3 to 8% iron, 0.15 to 0.30% phosphorus, and 2 to 8% silica. Most of it is mined from shallow workings.

In northeastern Tennessee, where the rocks are mainly limestone, sandstone, or quartzite, manganese is found in soil and clay, which is derived from the weathering of such rocks. A few of the deposits are in veinlike masses in the solid rock. Like in Arkansas, the ore is in small to large pockets. A carbonate is also found; it resembles dolomite, and is uncommon in this area.

Manganese oxide is mined from the oxidized ore-bodies of the Tintic region of Utah, particularly those mines that contain copper and gold.

Californian manganese is generally found in chert or jasper, which is frequently much folded and contorted. There are no large deposits in that State, save that in the Ladd mine near Livermore, where the lode is well-defined and the ore hard and of good grade.

At Butte, Montana, large quantities of rhodochrosite, the pink variety of manganese, is extracted from areas of zinc and silver-bearing ores. It must be concentrated before shipment. Philipsburg yields good ore. At Leadville, manganese is another of the many products of that district. It is either associated with silver or iron, and is mostly used as a flux at smelters. The grade runs from 10 to 35% manganese.

Throughout southern California, Nevada, and western Arizona are many deposits, one of the noted ones being near Las Vegas, in Nevada. The ore there is soft and light. Transportation is a factor against cheap operation in that arid region.

In the iron ranges of Lake Superior are great deposits of 5 to 10 and 10 to 35% manganiferous-iron ores, which have to be concentrated to marketable grade.

An excess of silica, above 10%, is not desirable, and

search should be made for ore low in this material. Many good-looking outcrops in jasper do not persist far below the surface.

The Russian deposits are beds in brown sandstone; in India they are in metamorphic rocks, inter-layered with quartzite; and in Brazil are chiefly the mineral psilomelane, the residual product of decomposition of an original manganese-bearing rock made up of the manganese carbonate and silicates.

Manganese ore in Cuba occurs principally in sedimentary rocks such as limestone, sandstone, and shale, that are in places metamorphosed (changed); and in beds that originally may have been water-laid tuff, but are now partly replaced by manganese oxide, zeolites, calcite, and other minerals. In the most heavily mineralized localities, the deposits are in and about masses of silicious rock — mostly jasper.

A field test for manganese is to powder some of the ore, mix an equal quantity of salt, place in a cup, and add sulphuric acid. Chlorine gas will be evolved, unmistakable on account of its pungent odor. The oxygen in the manganese combines with the acid to release the chlorine.

MOLYBDENITE

This is the sulphide. It occurs in crystallo-laminar masses or tabular crystals, having a strong metallic luster and lead-gray color, and forming a greenish-black streak which is best seen by drawing a piece across a china plate. It has a greasy feel. The flakes or scales resemble mica in the way they may be split. Specific gravity, 4.7 to 4.8; hardness, 1 to 1.5; easily scratched by the nail. It contains 59.95% molybdenum and 40.05% of sulphur. The other commercial molybdenum mineral is wulfenite, which is not very common and contains 26.15% molybdenum and 56% lead.

Geology of molybdenum. It occurs sparingly in granite, syenite, and chlorite-schists, and is sometimes mistaken for graphite, from which it is, however, readily distinguished by the streak, that of graphite being black, while the specific gravity of molybdenite is more than double. Near the Climax mine, near Leadville, Colorado, the largest molybdenite producer in the world, there were rich gold placers, and early prospectors noted the bluish and yellowish mineral. They thought it was graphite. A tunnel was driven for gold, and passed through a fractured silicified granite, in which the molybdenite was uniformly distributed. The mountain consists of alaskite, a highly silicious granite, an intrusion of the gneiss. The ore occurs as an intricate system of fine stringers through the alaskite. The original granite is much altered. The average content is 0.92% molybdenite, 2.7% iron, and a little copper.

At the Kingsgate mine, New South Wales, Australia, the molybdenite is found as small pipes with bismuthinite, making mining costly and milling difficult.

In Canada, especially in Quebec, molybdenite deposits are found in the Archæan regions, and are probably due to the influence of masses of granite. The ore occurs in quartz veins, pegmatite dikes (probably connected with the granite masses), and along contact borders of granite or pegmatite with crystalline granite. The ores carry from 1 to 3% molybdenite.

In California the mineral is widely distributed in quartz

and crystalline rocks, but in few places is. it segregated enough to work.

Tests. Before the blowpipe molybdenite is infusible, but tinges the flame faint green. Heated on charcoal for a long time it gives off a faint sulphurous odor and becomes encrusted white. Nitric acid decomposes the mineral, leaving a white or grayish residue.

A simple field test for molybdenum is as follows: Pick out carefully the supposed molybdenum mineral, grind fine, and take a quantity equal to a pea in size and put it in a glass test-tube with a speck of paper $\frac{1}{32}$ of an inch square. Add three drops of water and five drops of concentrated sulphuric acid, and boil slowly. If molybdenum is present the solution will change to a dirty green color when almost dry, and on cooling will turn a beautiful ultramarine blue. This color changes immediately on the addition of a few drops of water to a dirty gray. Care must be taken to take the pure molybdenum mineral, to boil the exact quantities specified, and to boil slowly. This test gives splendid results.

TITANIUM

Occurs in nature in the form of titanic oxide. The most important minerals containing titanium are:

Rutile. This mineral resembles cassiterite, but has a lower specific gravity, is red by transmitted light, and crystals are more prismatic. It occurs in striated prisms. many times twinned; in eight-sided prisms with flat pyramidal termination, rarely compact massive. Hardness — 6 to 6.5 ; specific gravity — 4.2 to 4.3 ; color — reddish-brown to black, by transmitted light deep red; streak —

yellowish-brown; and luster — adamantine. Occurs in granite, gneiss, syenite, mica-schist, and granular limestone. In North Carolina, rutile is abundant with corundum in creek gravels of Macon county; in South Carolina chiefly in gneiss; and in Virginia in pegmatite dikes, residual soil, and in gravels. It generally contains enough iron to give a reaction in the borax bead. With salt of phosphorus in reducing flame the strongly saturated bead is purplish-violet. It is used in making certain qualities of steel.

Other titanium minerals are octahedrite and brookite, which give similar reactions as rutile. Their mode of occurrence is frequently the same as rutile, and they, like it, often accompany gold. Their luster is adamantine, and they are likely to attract the eye when found in alluvial deposits, in which they are frequently associated with the diamond.

TUNGSTEN

During recent years this metal has become of great importance, due to its use in tool-steel and in electric lamps. The search for its minerals has been much accelerated since 1914. Tungsten is never found in native condition. It is very heavy, having a specific gravity of 19.3 to 21.4; and is very hard, ranging from 4.5 to 8. The important tungsten minerals are:

Scheelite. This is a tungstate of lime containing 80.6% tungstic trioxide or acid, and 19.4% lime. Hardness — 4.5 to 5; specific gravity — 5.9 to 6.1; color — white, yellow, and brown; and luster — greasy. Iron oxide discolors scheelite, rendering identification difficult.

There are four other minerals that rather confuse the prospector with scheelite, namely, barite, calcite, epidote, and garnet. The two latter are much harder; scheelite is easily scratched with a knife.

Search for tungsten deposits should be restricted to areas of quartzose — granitoid rocks. Scheelite is found as isolated crystals, as 'nuggets' in placer ground, and in quartz-feldspar veins. The second and third mentioned occur at Atolia, southern California, where is probably the largest scheelite mine in the world, and the deepest — 1200 feet. In the Tungsten hills, west of Bishop, Inyo county, California, are several important scheelite deposits, with mills. The ore is uncommon, being garnetiferous, lies in granitic rocks, and carries about 1% tungstic acid. Geologic conditions similar to those in the Tungsten hills prevail over a wide extent of country along the east slope of the Sierra Nevada. The places of contact of the intrusive granites with other rock, are the most likely places to prospect for other similar bodies of tungsten ore.

Scheelite frequently occurs with gold. Notable instances are the scheelite deposits in Otago, New Zealand, the Golden Chest mine in Idaho, the Union Hill mine in California, and the wolframite deposits at the Homestake and Wasp gold mines of South Dakota. Recovery of the gold is somewhat troublesome and has occasioned considerable research. It also occurs with silver, as at the Tip Top mine in Arizona, in White Pine county, Nevada, and in Bolivia, South America. Also with copper — a cupro-scheelite — in the great deposit at the Suan Concession in Korea, another metallurgical problem. There the occurrence is an extensive band of quartzite interbedded with limestone of an average thickness of 25 feet. Samples of several hundred thousand tons average 0.3% tungstic acid, 40 cents gold, and 0.4% copper.

Ferberite. This is a tungstate of iron, and deposits are mainly restricted to Boulder county, Colorado. It is black to light-brown in color, the latter due to hydrous iron oxide; its hardness is 5, and is easily scratched with a knife; and its specific gravity is 7.5. The cleavage is good but breaks other than with the cleavage showing a rough 'hackly' fracture. The streak is light-brown, due to limonite; but the powder from scratching with a knife is dark-brown to nearly black. Magnetite and ilmenite are mistaken for ferberite, but their streaks are different. Ferberite is found in veins cutting granite, pegmatite, and gneiss.

Hübnerite. This is a tungstate of manganese, mostly mined in White Pine county, Nevada. Its color is reddish-brown, caramel-brown, and black, with streaks brownish-red, light greenish-yellow, and somewhat greenish-yellow, respectively. Hardness — 5; specific gravity, 7.2, both characteristics almost similar to those of ferberite. Crystals are thin, flat, radial, and irregular. Garnet and sphalerite are sometimes mistaken for hübnerite; and manganese oxide coats it black.

Wolframite. This is a tungstate of iron and manganese, containing 76% tungstic trioxide, 16% iron, and 8% manganese. It is an important mineral, occurring extensively in South Dakota, Arizona, Bolivia, China, Burma, Portugal, and Australia. In the first named it is found in quartz veins with gold and tin. In Cornwall, wolframite, cassiterite, arseno-pyrite, and chalcopyrite are mined from a depth of 1800 feet. Its hardness is

5 to 5.5; and specific gravity 7.2 to 7.55. In these features it resembles the tin oxide, though somewhat softer also frequently in color, dark gray or brownish-red to brownish-black. The streak (or scratch powder), is a dark reddish-brown to black, luster is metallic to submetallic. The outward crystal form is rarely good, most wolframite occurring in irregular aggregates of individuals ranging from very small to 2 inches or more across. Hematite, manganese dioxide, magnetite, hornblende and other minerals are liable to confuse the prospector, but the above tests will guide him.

Other tungsten minerals of only mineralogical value are cupro-tungstite (copper and scheelite), tungstite (a hydrous oxide), tungstenite (a sulphide, Emma mine, Utah), and stolzite (a lead tungstate).

A very simple field test for tungsten ores is to treat a small quantity of the crushed material with hydrochloric acid in a test-tube. If the metal is present a canary yellow will be seen. To make sure of the test add a scrap of tin, which will give a blue coloration.

If heated in the oxidizing flame of a blowpipe with borax or phosphorus salt, all tungsten compounds form colorless beads. In the reducing flame they produce with borax a yellow glass, in case the quantity of tungsten is large. Small amounts added to the bead leave it colorless. When heated with a phosphorus salt in the reducing flame, a glass of pure blue color is produced, as long as tungsten compounds are present exclusively. On adding a particle of iron salt to the bead it assumes a blood-red coloration, and on introducing tin the bead becomes blue or bluish-green.

Another field test is to take 1 gram or 15½ grains of

crushed ore, mix it with a fusion powder containing potassium chlorate, potassium nitrate, sodium carbonate, and sugar, and apply a match. The result is a granular mass, which is rubbed a moment in an agate mortar, then placed in a test-tube with hydrochloric acid and sheet or granulated zinc. A little boiling will yield a blue-colored solution if tungsten be present. The equipment may be carried in a coat pocket and the test is easily made.

Materials needed for testing by blowpipe are 6 inches of platinum wire, a candle, blowpipe, 4 inch test-tubes, a bottle of concentrated hydrochloric acid, sodium carbonate, and granulated zinc.

URANIUM

These ores are comparatively rare. They occur generally in veins in the older rocks — granite, mica-schist, clay-slate, and porphyry. The most important mineral containing uranium is pitchblende or uraninite. This is a complex mineral containing rare earths, radium, lead, helium, nitrogen, and other elements. Color — pitchy-black to grayish-black; streak — brownish-black; hardness — 5.5; specific gravity — 9 to 9.7; and luster — pitch-like to dull. Crystals are rare, generally massive, compact, with a conchoidal to uneven fracture.

One of the principal sources of uranium is the carnotite ore of Colorado and Utah. This mineral is canary-yellow, containing also vanadium, lime, and potash. It is mined and treated for radium. In the States mentioned, carnotite is deposited in a light-colored, cross-bedded sandstone. It forms incrustations on exposures of the white sandstones. It is often granular. Float ore is sought, then traced to its source, which often takes considerable time. The orebodies are irregular and pockety. Tests for radio-activity are made by an electroscope.

Carnotite may be described as a complex ore consisting essentially of vanadium or with potassium as a double silicate, and associated with or loosely combined with uranium oxide. A ton of ordinary carnotite ore assaying 2½% uranium oxide contains only 12½ milligrams of radium.

It has been determined in Colorado and Utah, by the U. S. Bureau of Mines, that in well-sampled carnotite, the uranium content of which is accurately known, the radium content is not less than the proportion of 1 part radium to 3,000,000 parts of metallic uranium.

Hydrochloric acid does not attack pitchblende, but decomposes carbonates and silicates which may be present, as well as metallic sulphides, sulphuretted hydrogen being evolved. In a pure state pitchblende is soluble in nitric acid, the color of the solution being yellow. It dissolves with difficulty in concentrated sulphuric acid, the solution being green. The solution in nitric acid gives with ammonia a sulphur-yellow precipitate. When boiled with phosphoric acid an emerald-green solution results. The pure ore is next to infusible, colors borax yellow in the oxidizing flame, green in the reducing flame, while microcosmic salt gives a green color in both flames.

The element radium, which closely resembles barium in character, is extracted from pitchblende and uranium. Radium is used for medical purposes, and mixed with zinc sulphide forms a compound that glows in the dark. Other sources of uranium are from the pitchblende of Joachimstahl, Austria, and South Australia.

Pitchblende in Cornwall, England, occurs mostly as one of the minerals of copper deposits, though it is found with tin and other deposits. Fluorspar is also associated with it there (also at Jimtown in Colorado). Other uranium minerals, and compounds — probably the products of the alteration of pitchblende — are found with it. The pitchblende is likely to be coated with, or even entirely replaced by, alteration products; but at depth the black pitchblende appears, associated with sulphides and other unaltered minerals.

VANADIUM

This mineral is fairly widely distributed, but is not found anywhere in large masses. It does not occur native, but is found in Arizona, Colorado, and Utah in carnotite (see preceding page), roscoelite, and the following:

Vanadinite occurs in hexagonal prisms, with basal pinacoid, also in reniform aggregates of fine columnar to fibrous texture. Color — red to brownish-red; streak — white; luster — adamantine, on fracture resinous; hardness — 2.5 to 3; specific gravity — 6.6 to 7.23. Dissolves readily in nitric acid. Decrepitates strongly before the blowpipe, gives in the tube a slight white coating, fuses on charcoal to a globule which separates lead and gives a lead coating and produces in oxidizing flame of the borax globule a glass, red-yellow when warm, yellow-green when cold; and in the reducing flame a beautiful green glass. The ore occurs in lustrous red and yellow hexagonal crystals in Arizona; in green crystals with calcite at Charcas, Mexico, and with silver and lead ores in New Mexico, and in the Shattuck-Arizona mine, Arizona. Peru is the largest producer of vanadium in the world, contributing over 90%. The ore there is found as a red calcium vanadate (patronite) in pockets and fissures in fine shale. This mineral is greenish-black, contains from 19 to 24% vanadium oxide and 50 to 55% sulphur, has hardness of 2.5, and specific gravity of 2.71.

Descloizite. Pyramidal, resembling octahedrons, drusy. Color — olive-green to black; luster — adamantine to resinous; hardness — 3.5; specific gravity — 5.839. Gives water in a closed tube. The coating on charcoal reacts for zinc with cobalt nitrate. It occurs with lead and silver ores.

Dechenite. Color red to reddish-yellow; streak yellowish to orange. Transparent on fracture and edges. Occurs in lead ores of Leadville, Colorado.

Volborthite. This is a hydrous vanadate of copper, barium, and calcium. Hexagonal. Color — olive green, grass green, and yellow; hardness — 3 to 3.5; specific gravity — 3.45 to 3.55; streak — yellow. Gives water in a closed tube and becomes black. On charcoal a black slag is formed.

Vanadium during the past few years has become of great importance to the steel trade. Its alloy with iron is known as vanadium steel, and is highly valuable in the construction of automobiles, locomotives, and the like.

Special works on preceding metals:

Chromite — By S. H. Dolbear and Albert Burch, San Francisco, 1918. Bulletin of Ontario Bureau of Mines, Toronto.

Manganese — U. S. Geological Survey and Bureau of
    Mines.
Molybdenite — Bulletin 111 of U. S. Bureau of Mines, by
    F. W. Horton, 1916.
Tungsten — Bulletin 12 of South Dakota School of
    Mines, by J. J. Runner and M. L. Hartmann, 1918,
    Bulletin 652 of U. S. Geological Survey, by F. L.
    Hess, 1917.
Uranium — Bulletin 103 of U. S. Bureau of Mines, by
    K. L. Kithil and J. A. Davis, 1917.
and the current technical press.

CHAPTER XI

TIN

This is an important metal and hardly replaceable by any other. Its principal uses are in making tin-plate, solder, and composition metals. Metallic tin is rare in nature. Its hardness is 2, and specific gravity 7.18. The world's annual output is around 100,000 tons, produced in the Malay States, Bolivia, Dutch East Indies, Australia, Cornwall, and China.

Cassiterite or Tin Stone. This mineral is the principal source of tin, and when pure contains 78.6% of metal. It is remarkable for its hardness — 6 to 7 — and still more so for its specific gravity — 6.8 to 7. It contains small quantities of iron, copper, manganese, tungsten, tantalite, arsenic, sometimes silica, and rarely lime. It is found associated with quartz, mica, topaz, tourmaline, wolfram (as in South Dakota), chlorite, iron, copper (as in Cornwall), silver (as in Bolivia), and arsenical pyrite. It occurs massive and in crystals, also in botryoidal and reniform shapes, concentric in structure and radiated fibrous, and is then in the last form called ' wood tin,' from its woody appearance. ' Toad-eye ' tin is the last described, but in small shot-like grains. Stream tin is nothing but ore about the size of sand, found along the beds of a stream or in the gravel of the adjoining region. Its source was tin veins or rocks. Two of the

223

great tin mines of Tasmania were gravel mines. The stream tin lay under an overburden of basalt.

Cassiterite yields a white, grayish, or brownish streak; has a brownish color and a dull luster. It is nearly as hard as quartz, and will scratch glass, especially if freshly broken. Pure crystals are rare, and are nearly transparent. Generally the ore is of a dark-brown color, and sometimes almost black; the fine powder or streak made by a file is light brown, however dark the mineral may be. The brown shade is due to oxide of iron mixed with it; if perfectly free from impurities it would be nearly white or colorless. The usual appearance in mass or pebbles, or finer, is that of a dirty or burned-brown color with varying depths of shade.

In the gravel form it is inclined to wear smooth, due to its extreme hardness. It was in this form that it was discovered in Banca and Billiton (Dutch East Indies) in 1710 and traced to its source in the hills, where the central rock is granite, covered by quartzites, altered sandstones, and slate. The altered sandstone just above the granite is the most productive rock, and it is traversed in all directions by tourmaline.

Another tin mineral, of little commercial value, that may lead to the discovery of cassiterite is tin pyrite, the sulphide of tin. Its composition is 29 to 30% sulphur, 25 to 31% tin, 29 to 30% copper, with iron and sometimes zinc. It has been mined as an ore of copper and called 'bell-metal.' Its hardness is 4; specific gravity 4.3 to 4.5; has a metallic luster; color, steel-gray to black, often yellowish from the presence of copper sulphide; and is opaque and brittle.

With nitric acid it gives a blue solution; the sulphur

and tin oxide separate, and may be tested on charcoal, where it fuses to a globule, which, in the oxidizing flame, gives off sulphur and coats the coal with white oxide of tin.

In the United States, cassiterite occurs in small stringers and veins on the borders of granite knobs or bosses, either in the granite itself or in the adjacent rocks, in such relations that it is doubtless the result of fumarole action consequent on the intrusion of the granite. It appears that the tin oxide has probably been formed from the fluoride. The South Dakotan deposits are in pegmatites and quartz, while some stream tin has been worked. At Broad Arrow, near Ashland, Alabama, tin is disseminated in gneiss, the ore averaging 1½% black tin (the oxide) but is much mixed with titaniferous iron. At King's mountain, North Carolina, cassiterite occurs irregularly in a greisen or altered granite, and in limited alluvials derived from the disintegration of the same. On Irish creek, Virginia, test lots taken from deposits in granite have yielded up to 3½% metallic tin, largely associated with arsenical pyrite and ilmenite, which increase the difficulties of concentration and lower the value of the product.

On the Seward peninsula, Alaska, stream tin has been found in the sluice-boxes for saving gold on Buhner creek and on the Anikovik river. At the former place, 2 to 3 ft. of gravel overlies the bedrock, which consists of arinaceous schists — often graphitic — together with some graphitic slates. Bedrock is much jointed, the schists being broken up into pencil-shaped fragments. They strike nearly at right angles to the course of the stream and offer natural riffles for the concentration of heavier material. The slates and schists are everywhere penetrated by small veins, consisting usually of quartz with some calcite and frequently carrying pyrite and sometimes gold. These veins are very irregular, often widening out to form ' blebs,' and again contracting so as not to be easily traceable. The cassiterite occurs as grains and pebbles, from those microscopic in size to those half an inch in diameter, and varies in color from a light brown to a lustrous black. During the past few years, two or three dredges have been recovering black tin from York peninsula, west-central part of Seward peninsula, the total annual yield being around 100 tons. This is really the only tin mining in America.

Tin occurs in El Paso county, Texas. The cassiterite carries wolframite (tungstate of iron and manganese), and is found in aplite dikes cutting granite. Little has been heard of the deposits in recent years. Other occurrences are in Riverside county, California, where the country rock is granite and the vein is tourmaline and quartz; and in Spokane county, Washington, in pegmatite cutting gneiss, quartzites, and schists. While some of the American deposits have possibilities, few people are bothering about them, yet the development of tin in this country is very important as it consumes 40% of the world's output and produces almost none.

Cassiterite stands nearly by itself in its mode of occurrence and formation, as a type of a strongly marked class of deposits. It is always associated with granitic rocks, quartz-porphyries, or gneiss, all of which are of analogous composition, rich in silica, which crystallizes as quartz, and are consequently called acid rocks. Tin lodes are nearly all of great age, and occur only in those of the above-named rocks that are characterized by the presence of white mica. Cassiterite rarely occurs in green rocks, unless their color be due to chlorite; nor in dark-colored rocks, except where stained red by the decomposition of iron minerals; neither is it found in limestone.

Those granites that are characterized by abundance of white mica have, with good reason, been termed 'tin granites,' and a coarse-grained rock composed of granular quartz mixed with this mica and called 'greisen' occurs in all the tin fields of the world.

Tin is mined from the surface, such as at Mt. Bischoff, Tasmania, or at great depths, such as 3000 ft. in the Dolcoath mine in Cornwall.

Tests. When a tin-bearing mineral is heated before a blowpipe with carbonate of soda or charcoal, while metallic tin separates out. By dissolving this in hydrochloric acid and adding metallic zinc, the tin will be deposited in spongy form. In the blowpipe assay, tin leaves behind a white deposit, which cannot be driven off in either flame. If it be moistened with nitrate of cobalt solution, the deposit becomes bluish-green, this test distinguishing it from other metals.

If the ore is poor it should be concentrated in order to remove the gangue or worthless rock. If mixed with iron or copper pyrite, it should be calcined or else treated with acid. One method is to mix the ore with one-fifth of its weight of anthracite coal or charcoal, and melt it in a crucible for 20 minutes. The contents are then poured into an iron mold, and the slag carefully examined for buttons or prills of tin.

Another method is to mix 100 grains of the ore with six times its weight of cyanide of potassium, and melt it for 20 minutes. The contents are either poured into a mold or allowed to cool, after which the crucible is broken to separate the slag and tin button. As cyanide is a deadly poison, care should be exercised in handling it and keeping clear of the fumes.

CHAPTER XII

MERCURY, BISMUTH, NICKEL, COBALT, AND CADMIUM

Mercury or Quicksilver. This is the only metal that is liquid at ordinary temperatures. Its color is tin-white. The usual properties of a metal are, however, highly developed in it, and when solid it has much resemblance to silver, especially in its high metallic luster, ductility, malleability, its capability of being cut with a knife, its granular fracture, and its high degree of conductibility of heat and electricity.

Mercury readily combines with most of the other metals, and the compounds thus formed are called 'amalgams.' Such combinations with the heavy elements are generally easy of decomposition, hence it is exceedingly useful for the recovery of gold and silver from ores. The quicksilver picks up the almost invisible specks of gold, and in this way the gold is concentrated into a comparatively small space. By heating the amalgam, the mercury is driven off, the fumes condensed and the metal saved, while the gold is separated in a form ready for melting into bars.

Native Mercury. This occurs occasionally as globules disseminated through the rocks. Most cinnabar mines show this, it being formed either by reduction of the sulphide or by sublimation of mercuric vapors. Color is tin-white; specific gravity — 13.6 at 50° F., and about 15.6 when solid.

Sometimes native amalgams of gold, silver, and mercury are found. The one most frequently found is with silver, which, when pure, contains from 64 to 72% mercury. Color is silver-white; hardness — 3 to 3.5; and specific gravity — 10.5 to 14. It is soluble in nitric acid. On charcoal before the blowpipe, the mercury evaporates or volatilizes, and the silver remains. Amalgams of gold and mercury have been found in Utah; also in Chile and Norway. The rich silver ores of Cobalt, Ontario, carry a good deal of quicksilver, but not as an amalgam.

Cinnabar or Sulphide of Mercury. The main supply of commercial quicksilver is obtained from this ore. It is found in granular, fibrous, dense and earthy masses, and sometimes also in small rhombohedral or prismatic crystals. Specific gravity — 6.7 to 8.2; hardness — 2 to 2.5; streak, scarlet-red; fracture uneven, splintery; luster, adamantine; and color, generally cochineal-red; also brown, brownish-black, etc. It contains 86.2% mercury and 13.8% sulphur, when pure. Before the blowpipe in a closed tube it yields a black sublimate of the same composition as the original mineral. In the open tube, if carefully heated, it yields sulphurous acid and mercury globules, together with a small quantity of black sublimate. On charcoal it volatilizes completely. It is insoluble in sulphuric acid, hydrochloric acid and potash lye; but soluble in aqua regia, sulphur being separated; also in solutions of potassium or sodium cyanide, and sodium sulphide with sodium potash.

Other, though non-commercial minerals are meta-cinnabarite, calomel, coloradoite, tremannite, guadal-cazarite, and kleinite.

The quicksilver deposits of Almaden, Spain, have a far remote history, and occur in upper-Silurian slates, sometimes inter-stratified with beds of limestone; but the ordinary slates themselves, which are much contorted, rarely contain cinnabar. The enclosing rock usually consists of black carbonaceous slates and quartzites alternating with schists and fine-grained sandstones.

At Idria, Austria, cinnabar is found in impregnated beds and stockwerks, in bituminous shales, dolomitic sandstones and limestone breccias of Triassic age, dipping 30 to 40°, and covered by carboniferous sandstones and shales in a reversed position. This deposit has been worked for nearly 400 years, and is said to become richer with depth.

The quicksilver-bearing belt of California extends along the coast range for a distance of about 200 miles. These deposits are generally impregnations in Cretaceous and Tertiary formations. They seem to be richer when the beds are more schistose and transmuted. They are more or less closely in relation with serpentines, which are themselves sometimes impregnated with oxide of iron, sometimes in quartzose schists, in sandstone, more rarely in limestone rocks, limestone breccias, etc. In general, the Californian deposits occur along the contact between serpentine and metamorphic sandstones and shales, and the mineral has been deposited from solfotaric waters carrying the sulphide in solution. These solutions impregnated the sandstones and brecciated masses of opal and chalcedony that have formed in the serpentine through much silicification, leaving seams and pockets of cinnabar.

Native mercury is found in some magnesian rocks near the surface. There are no defined fissures nor veins proper. The cinnabar, with quartz, pyrite, and bituminous substances, is sometimes disseminated in the rock in fine particles and spots, sometimes forms certain kinds of stockwerks or reticulated veins and nests. The parts thus impregnated congregate and form rich zones, carrying 35% metal. These rich zones without defined limits gradually merge into poor stuff containing half a per cent., or mere traces, and are of no value.

Quicksilver deposits are not deep-seated anywhere, and those worked are not rich, averaging under 1% metal in California. Eight miles southeast of Mina, Nevada, is the Pilot range, in which is a belt of cinnabar deposits worth exploration. In the Goldbank mine, near Winnenucca, Nevada, cinnabar is disseminated in silica of recemented breccia.

In the Mazatzal range in central Arizona, the lodes consist of veinlets, films, and specks of cinnabar in schist, and as a rule have no definite walls. Gangue minerals below the zone of oxidation are calcite, limonite, and quartz.

In the Terlingua district of Texas, quicksilver ore is confined to limestones and shales. The region is arid and far from rail, yet is worked profitably.

Cinnabar cannot be mistaken in the field, its color and streak are typical of this mineral alone. With care, the blowpipe will ' retort ' some mercury from it which may be left on charcoal as globules. In the Idria district of California, the serpentine belt is characterized by scarcity of underbrush and sparseness of the timber growth. Chromite float and sand is found in creeks and originally came from this serpentine. All districts in which are hot springs or sulphur deposits, should be examined for cinnabar. An instance of this is the Sulphur Bank mine in Lake county, California, which was originally worked for sulphur, but later on developed into an important quicksilver producer. The accepted theory of origin of cinnabar is one of hydro-thermal deposition, hot springs being the agencies, either accompanied or followed immediately after periods of volcanic activity or other igneous intrusions.

BISMUTH

This metal occurs native, and has a silver-white color with reddish tinge, tarnishing dark brown. It is brittle when cold; hardness, 2 to 2.5; specific gravity, 9.7 to 9.8. It is malleable and sectile when heated, but breaks under the hammer. It carries, sometimes, traces of arsenic, sulphur, tellurium, and iron. On charcoal before the blowpipe, bismuth fuses and entirely volatilizes, leaving a coating which is orange-yellow while hot and lemon-yellow on cooling; this is the trioxide of bismuth. It dissolves in nitric acid, but subsequent dilution causes a white precipitate.

Very little bismuth has been found in America. About 100 pounds was once found in a pegmatite vein of quartz, feldspar, tourmaline, etc., in San Diego county, California. Where it has been found in the United States it has been associated with wolfram, galena, and zinc blende in quartz. Bismuth is one of the products of the electrolytic refining of copper and lead. The metal occurs in Europe, associated with silver and cobalt, also with copper ores. The principal uses of bismuth are in alloys of low melting point, and in drugs.

The geology of bismuth is the same as that of copper; it occurs in veins in gneiss and other crystalline rocks and clay slate, accompanying ores of silver, copper, lead and zinc.

Bismuthinite. This is the sulphide, containing 81.2% metal. It has a lead-gray color, metallic luster, hardness of 2, and specific gravity of 6.4. It is found in small quantities in Colorado, Utah, and Washington. In New South Wales, Australia, there are bismuthinite-molybdenite deposits, also in Tasmania.

NICKEL

This metal does not occur native except in meteorites.

Under the blowpipe, the test for nickel requires care and some practice. On charcoal, with soda in the inner flame, it gives a gray metallic powder, attractable by the magnet. In the outer flame the borax bead shows a hyacinth-red to violet-brown while hot, a yellowish or yellow-red when cold. In the reducing or inner flame, a gray appearance is given. These results are modified by impurities and the amount of nickel in the mineral. The wet process is the only method of determining the true value of a nickel-bearing mineral. The chief ores are:

Smaltite, which is a combination of cobalt, iron, and nickel, and arsenic in varying proportion. It will be more fully referred to under 'cobalt.'

Nickel Arsenide, Copper Nickel or Niccolite. The composition is nickel, 44.1%; and arsenic, 55.9%. It looks somewhat like pale copper, but contains no copper.

Its hardness is 5 to 5.5; specific gravity, 7.2 to 7.8; streak, pale brownish to black; and luster, metallic. It is brittle. Niccolite frequently contains a little iron, and sometimes a trace of antimony, lead and cobalt. It dissolves completely in aqua regia, also in nitric acid, crystals of arsenious acid being separated by cooling.

If carefully treated under the blowpipe with borax, niccolite will show the iron, if present, in the bead, and the cobalt and nickel by successive oxidations (see under smaltite).

There is another mineral, not properly an ore, called emerald-nickel, a carbonate of nickel, containing 28.6% water when pure. It forms incrustations on other minerals, like another

Millerite, a sulphide, carrying 64.1% nickel. This forms tufts of fine acicular, brassy-looking crystals, in cavities of red hematite in the Sterling iron mines in northern New York, and velvety incrustations on copper ores in the Gap mine, Lancaster county, Pennsylvania, where nickel was found and worked. In the former place there was no nickel in quantity, but in the latter it has in the past been found in commercial amounts, but the mine is now exhausted. The sulphide forms at the Gap varied very much, when examined under the microscope, from the acicular crystals found in the ores at Sterling, yet they are of the same chemical combination. The ore upon which the tufts of velvety covering were found at the Gap is pyrrhotite or sulphide of iron, holding 4 to 5.9% nickel in composition.

The deposits of nickel discovered in Sudbury, Ontario, north of Georgian bay, yield nickel in pyrrhotite (sulphide of iron), and apparently also in chalcopyrite, whose typical composition is copper, 34.6%; iron, 30.5%, and sulphur, 34.9%. It is a mineral of brass-yellow appearance, and one that carries the copper of Cornwall, England, and at the copper beds in Fahlun, Sweden. In the latter place it is imbedded, as it appears to be in the Sudbury region, only that the ore is imbedded in pyrrhotite and the Swedish in gneiss.

The chalcopyrite does not mix intimately with the nickel ore so as to form a homogeneous mass; it occurs by itself in pockets or threads, but enclosed with massive pyrrhotite, which, while it may have more than 30% of nickel present, does not show any sign of the changed composition.

This quantity is far above the average of nickel in the pyrrhotite, which seldom carries less than 2½% or more than 9% of nickel.

The following minerals exist at Sudbury:

Foleyrite, of a bronze-yellow color, grayish-black streak, and metallic luster. It occurs massive and contains 32.87% of nickel. Its specific gravity is 4.73, and hardness, 3.5.

Whartonite contains 6.10% of nickel. It has a pale bronze-yellow color, black streak, and metallic luster. Specific gravity is 3.73; and hardness, 4.

Jack's Tin or Blueite contains 3.5% of nickel. It is of an olive-gray to bronze color, metallic luster and black streak. Specific gravity is 4.2; and hardness, 3 to 3.5.

In addition to the nickel ores already given may be mentioned a hydrated silicate of nickel found in New Caledonia and named after its discoverer:

Garnierite. It contains from 8 to 10% nickel. Color is light or dark green; streak, light green; specific gravity,

2.2 to 2.86; and hardness, 2.5. It fuses in borax before the reducing flame, and gives the ordinary nickel bead. It is found in lodes and pockets in serpentine rock. It has also been found in Oregon.

Nickel has been found as ore in Douglas county, and as sand in Josephine county, Oregon, but not in payable quantities.

The nickel-copper deposits at Sudbury, Ontario, occur mostly along the outer, basic, contact of the norite (a silicate of alumina) micro-pegmatite, but the commercial orebodies are rarely found in the norite. In the Creighton mine, ore is at the granite-norite contact, mostly in the granite footwall. Sudbury ore is accompanied by rock fragments, from specks up to 15 feet diameter. This renders sorting necessary. The Norway formations are similar to those at Sudbury.

At the Alexo mine, a considerable distance from Sudbury, the lode is in a hollow-lava (andesite-rhyolite) and serpentine contact. The ore is both massive and disseminated.

In New Caledonia, near Australia, the deposits are vein-like, brecciated, and masses of altered serpentine impregnated with nickel; also nickel earths. The Cuban deposits are similar to those of New Caledonia, and have been formed by weathering of the basic rock, now a serpentine, and rest on the surface of the rock from which they have been derived.

Nickel is also found in the iron ore of Cuba, in the manganese ore wad, in blister copper, and in silver-cobalt ore.

COBALT

Cobalt does not occur in native form. The following are the minerals of importance:

Smaltite is composed of cobalt, nickel, iron, and arsenic; the typical form is arsenic, 72.1% ; cobalt, 9.4% ; nickel, 9.5% ; iron, 9%. When pure it contains 28.2% cobalt. Hardness is 5.5 to 6; specific gravity, 6.4 to 6.6; color, tin-white, sometimes iridescent; streak, grayish-black; luster, metallic; and it is brittle. Before the blow-pipe, on charcoal with soda, the arsenious acid fumes are given off, and the garlic smell is plain. With borax for the bead the assay may be made to show (with successive heatings) the reactions, first of iron, then cobalt, and nickel, provided the operator is skilful in oxidizing the powdered ore cautiously by degrees. When one borax bead shows iron reaction by a certain amount of carefully applied oxidizing flame to the bead, try another with increased degree of oxidation until the cobalt blue and nickel brown are noted if both are present.

Through replacement of the cobalt by nickel, smaltite grades into chloanthite, the nickel arsenide. The most abundant deposits of smaltite are at Cobalt, Ontario, where the ore carries cobalt, nickel, arsenic, silver, and mercury. The ore is treated for silver, and cobalt is saved during treatment of concentrates at smelters.

Cobaltite is composed of sulphur, arsenic, and cobalt in the typical proportions of 19.3, 45.2 and 35.5% ; but it frequently contains iron. Its hardness is 5.5 ; and specific gravity, 6 to 6.3. Under the blowpipe, in an open tube, it gives off sulphurous fumes and a sublimate of arsenious acid. With borax, the bead gives cobalt blue. It dissolves in warm nitric acid, separating the sulphur and arsenic.

Cobaltite and smaltite are valuable as affording the greater part of the smalt of commerce, and the former is used in porcelain painting, giving a deep blue color. Cobalt is now used for enameling metal-ware, giving a similar finish to that of nickel.

Erythrite is a soft (hardness 1.5 to 2.5) peach-red mineral with a specific gravity of 2.9, transparent or translucent, sometimes pearl or greenish-gray. Its composition is arsenic, 38.43%; cobalt oxide, 37.55%, and water, 24.02%.

Erythrite is found in the oxidized parts of cobalt and arsenic-bearing veins.

In a closed tube, under blowpipe, it yields water and turns bluish. It gives the usual blue for cobalt in the borax bead.

Linnæite. A sulphide of cobalt, is valuable for the large amount of both cobalt and nickel it sometimes contains. Hardness, 5.5; specific gravity, 4.8 to 5; metallic luster; color, pale steel-gray, tarnishing to red. Composition, sulphur, 42%; cobalt, 58%, but cobalt is replaced by large amounts of nickel, and sometimes copper. Some specimens from Mineral Hill, Maryland, and from Missouri, have yielded as high as 29.56 and 30% of nickel, with 21 to 25% of cobalt in the same specimen, but with a small amount of iron — 3%.

Asbolite is an earthy cobalt or cobalt wad occurring as a bog ore, with manganese, iron, and copper, and nickel. It is blue, black at times, has a hardness of 1 to 1.5, and specific gravity of 2.2 to 2.6. It sometimes contains up to 35% of cobalt oxide.

The geological position of cobalt is in the earlier rocks, as the chlorite slates with chalcopyrite, blende, and pyrite, as in Maryland. Sometimes the ore is found in cavities in the limestone of the Carboniferous age, as in Great Britain. The tin-white cobalt is found in the gneissic and primitive rocks as in Norway. Linnæite is found at Mine la Motte, Missouri, in bunches and masses, sometimes in octahedral crystals among its rich ores of lead and nickel.

CADMIUM

Of this mineral but one ore is known, namely, the sulphide, or 'greenockite,' containing 77.7% cadmium. Color, honey to orange-yellow and brick-red ; in hexagonal prisms; hardness, 3 to 3.5; specific gravity, 4.9 to 5.0. Before the blowpipe, on charcoal with soda, it yields a red-brown deposit. Cadmium is frequently associated with zinc ores, some varieties of sphalerite or blende containing 3.4%.

Metallic cadmium is white like tin, and shares with it the property of emitting a crackling sound when bent, and has been used successfully in place of tin in solder. It is so soft that it leaves a mark upon paper. Most cadmium is a by-product of the electrolytic refining of base metals.

CHAPTER XIII

ALUMINUM

Distribution of aluminum in nature is extensive, rivaling that of iron, yet there are few minerals that serve as sources of the metal. All clays contain a large percentage of aluminum, but always in the state of silicate, and the difficulty of removing this silica has so far prevented the employment of clay as an ore of aluminum.

Bauxite. This mineral, which derives its name from Baux in France, is the principal source of aluminum, alum salts, abrasives and refractories. The chief deposits in America are in Alabama, Arkansas, and Georgia. In the first State, search is confined to edges of red clay areas adjacent to quartzite ridges. Pebbles of bauxite are an indication of deposits. Some of the ore is found as pockets, others as blanket deposits.

The Alabaman deposits all occur in the lower Silurian formation. The district has been badly broken up by sharp folds and great thrust-faults, and the mineral occurs as pockets in close association with brown iron ore (limonite) and clay.

Bauxite is a limonite in which most of the iron is replaced by aluminum. It occurs in earthy masses resembling clay; also in compact form. The color varies from white to gray, yellow, also to brown or red, especially in the impurer kinds. It contains 50 to 75% alumina. Its hardness is 1.5, and specific gravity, 2.5. Corundum, generally pure, is too valuable for abrasive purposes to be used as ah ore. It will be more fully described in Chapter XVI, this mineral species including some of the most important precious stones. Ore should contain at least 20% aluminum before being offered to buyers.

Cryolite, of which there is practically only one productive mine, that at Ivigtut, southern Greenland. The mine is worked as a quarry, and has been extensively developed. The vein appears to widen with depth, but the quality of the mineral becomes inferior. Several thousand tons of cryolite annually are shipped to the United States.

Its specific gravity is 2.9. It contains 40% aluminum fluoride and 60% sodium fluoride.

With the blowpipe, on charcoal, cryolite fuses to a clear bead, becoming opaque on cooling. After long blowing with the oxidizing flame the assay spreads out, the fluoride of sodium sinks into the charcoal, and the suffocating odor of fluorine is given off, and the aluminum remains as a crust which, if touched with a little cobalt solution and gently heated, gives a blue color of alumina. If some of the cryolite be powdered and placed near the open end of a glass tube and the flame from the blowpipe carefully turned upon it, the fluorine will be freed and will etch the glass, showing corrosion and proving the presence of fluorine. Besides, as a source of the metal aluminum, cryolite is used as a flux, and largely for the manufacture of white enamel for iron ware.

While the older processes of aluminum manufacture, dependent on the reduction of the double chloride of aluminum and sodium, must always have a scientific interest, they have been beaten out of the field of commercial industry by the newer methods using electric furnaces, of which there are four varieties: In England and America, Cowles' and Hall's patents are followed; on the Continent, Heroult's and Minet's. They are all virtually modifications of the original Deville-Bunsen process, maintaining fusion by the heat of an electric current.

CHAPTER XIV

PETROLEUM, ASPHALT, OIL-SHALE, OZOKERITE

PETROLEUM

There is at the present time, and is likely to be indefinitely, a world-wide interest in oil. This was induced by the events of the past five years, calling attention to the need of oil and its products for all purposes, especially gasoline for automobiles, airplanes, tractors, trucks, and motor-boats; also fuel-oil for steamers, metallurgical works, and Diesel engines; and lubricants for all machinery. Development of oil-lands in North and South America is active, and prices are high. The principal centers of interest are Texas and Wyoming, but prospecting is under way in Arizona, California, Louisiana, Montana, New Mexico, and Utah. The United States' production of crude oil in 1919 was 376,000,000 barrels, the highest yield for any one year. Oklahoma was first, followed by California.

It might be said that prospecting for oil is a special study. The U. S. Geological Survey considers that in predicting the possible discovery of oil and gas in any region the method now generally employed, with great success, is to examine thoroughly every feature of the geology of the region, especially the features that determine or affect the origin and accumulation of oil and gas, and next to compare carefully the features thus disclosed with the features of the nearest fields in which oil is obtained from the formations found in the field just explored. Conclusions based on work of this kind done by thoroughgoing and competent geologists in many parts of the country score more than 60% in the discovery of new oil and gas pools. This is roughly ten to one better than the old-style methods, which took little account of certain geologic details now known to be very significant; and as geologic investigation and experience in the oil-fields go forward the percentage of success is increasing. Nevertheless, the drill must give the final answer to the question whether oil or gas will be found at any specific point in ' wildcat ' territory.

Some of the possible oil-lands are in Pike county, southwestern Arkansas; in the Salinas Valley-Parkfield district of California; in Teton and Lewis and Clark counties of northwestern Montana, also in south-central Montana; in east-central New Mexico; in the Nesson anticline of Williams county, North Dakota; in northeastern Texas and southern Oklahoma; and in Carbon county, Utah, where there is sandstone saturated with asphalt, and certain oil seeps suggest that there may be an oil accumulation nearby.

There are two theories as to the origin of petroleum — the inorganic and the organic. Both are based on careful study, but it seems that the generally accepted theory' is the organic. Any organic substance that becomes enclosed within sedimentary rocks may be a source of oil. Pressure, temperature, and the action of circulating waters are all factors in its production. Then a receptacle must be found for accumulations of the oil, so sandstone, limestone, and shale are the porous media, overlain by an impervious cover or rock.   Oil-bearing formations usually consist of alternating beds of sand and sandstone, clay or shale, and limestone.   The shale, considered to be the main source of the oil, which then migrated to the sands, is generally twice as thick as the limestone. A region in which the strata contains a small quantity of sandstone, a large amount of shale, with some limestone, there should be a chance of finding oil.   A surficial study of the formation is necessary, although some oil-pools can only be found by drilling.   The presence of oil in these rocks is indicated at the surface by oil-seeps or springs, gas emanations, asphalt deposits, or rocks giving an oily odor.   Such indications do not always mean large quantities of oil, as is proved by the large sums of money that have been spent in exploiting favorable looking formations.   A lack of surface indications does not mean that there is no oil in the rocks, as many important producing areas showed nothing to indicate that they contained petroleum.

In Oklahoma, the largest producer of oil in America, prospecting instruments are not to be relied upon, and in the location of wells a knowledge of the geologic conditions and structural features is necessary.   Showings of oil and gas may be found in any of the sedimentary rocks. Asphalt and oil seepages often serve as aids in the proper site for test wells.

In discussing the source of oil in Oklahoma, C. L. Snider considers that it is almost undoubtedly from plant or animal matter, or both, which was buried with the rocks as they accumulated at the bottom of the sea.   As the muds and sands were hardened by pressure, and finally elevated to their present position, the organic matter in them, being shut off from the air, was subjected to a process of decay, and the series of compounds forming natural gas and petroleum resulted. At first these compounds were distributed throughout the rocks in extremely small particles. Later, the folding of the rocks

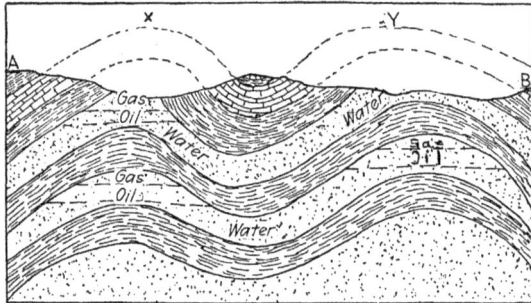

Fig. 54.

In this figure the sandstone is represented by the dotted pattern, and the shale by the closely-ruled pattern. The line A-B represents the present land surface, and the broken lines represent the strata that have been removed by erosion. The oil and gas are shown collected in the sandstones under the crests of the anticlines at X and Y. At Y the shale has been removed from over the upper sandstone, allowing the oil and gas to escape.

created places in the arches of the folds where the gas and oil particles and that of the salt water with which they were associated, accumulated. The former, being lighter than water, collected in the highest places. These are the crests of arches of the gentle folds that are present in the rocks of almost all regions. To accomplish this accumulation, there must be above the porous rock containing the gas and petroleum a tight-grained rock through which they cannot pass. This rock is in most places a shale and is called a ' cap-rock.' The arches or crests of the folds in which gas and oil may be collected are called anticlines, and the troughs between the ' anticlines ' are called ' synclines.' The porous rock in which the gas and oil collect is the ' reservoir rock ' or ' oil-sand.' In general the gas accumulates in the tops of the folds, that is, along the crests of the anticlines; the petroleum is found farther down the slope; and the salt water still

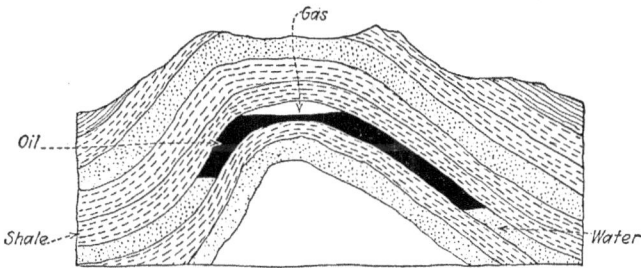

FIG. 54A.— AN ANTICLINE, SHOWING WHERE GAS, OIL AND WATER ACCUMULATE.

farther down the slope, and in the synclines. This regular arrangement may be greatly complicated by the difference in thickness and porosity of the oil-sand, and other factors.

Crude petroleum occurs only in the higher strata of rocks, it being never found in metamorphic rocks or crystalline formation. The Pennsylvanian oil strata belong to the Devonian age, the anticlinal ridges being more favorable, it is said, than the synclinal ones. Sandstones saturated with oil form the reservoir. This rock appears to be lenticular in form, and of varying texture, sometimes passing into conglomerates. The following deductions have been ascertained from the Pennsylvanian region: (1) The thicker the cover the more the oil, large accumulations being seldom found under light covers; (2) the coarser and more open the sand the more the oil; (3) the sandstones buried in shales must form the reservoir; and (4) underlying shales must exist, as they are the source of the oil.

In the Ohio region the Trenton limestone, which is struck at a depth of from 1100 to 2200 feet, and is covered by 400 to 1000 feet of shale, appears to be both the producer and reservoir. The principal accumulation both of oil and gas, are always the uppermost beds of the limestones, and generally not more than 20 or 30 feet below its upper surface. The oil-rock continues to a lower level, but below the oil it is charged with brine containing unusual quantities of chloride of calcium and magnesium. When this is struck the well is frequently lost, although it is sometimes possible to plug it near the bottom. The problem of water in oil-wells is a serious one, and in California the State Mining Bureau and oil companies co-operate to combat the trouble. The limestone is porous in parts, but this seems to be due to dolomitization, the change having resulted in re-crystallization, which has left innumerable microscopic cavities, in which the oil has accumulated.

In Kentucky, petroleum occurs near the base of carboniferous limestone; in California, it is found in strata belonging to the Tertiary age, shales and sandstones being the source of a large part of the oil; in Colorado and other western States, in those belonging to the Cretaceous; and in North Carolina, to those belonging to the

Triassic. In West Virginia oil is found in strata belonging to the coal measures. The Gulf coastal plain oilfield of Texas includes a belt of country from 50 to 75 miles wide, bordering the Gulf of Mexico, and extending from the vicinity of the Mississippi River in Louisiana westward about two-thirds of the distance across Texas. The oil-rock is a crystalline dolomitic limestone, having an extremely porous structure; associated with it is considerable selenite or crystalline gypsum. Another abundant crystalline accessory mineral is native sulphur. In north-central Texas, where there was such an oil-boom in 1919, the oil-pools lie in a Carboniferous belt of rocks consisting of sandstones, limestones, and shales. The producing sands lie at depths of 200 to 3150 feet. All of the pools discovered occur on a fold of the Carboniferous beds.

Crude petroleum is a fluid of a dark ‚color, sometimes black, and contains 84 to 88% of carbon, the remainder being hydrogen. Oil found in different areas vary widely in composition and appearance. Even in the same field, some wells yield a heavy, thick, black oil of low gravity, while others give a thin high-gravity oil. Payment at the wells is made on a gravity basis, starting at about 14 degrees Baume scale. This gravity should not be confused with specific gravity as applied to rocks.

In looking for petroleum, the prospector, besides the customary outfit, should carry a stick provided with a long iron point. It is best to follow water-courses upward, because progress will not be impeded by the turbidity of the water. It is also advisable to make such excursions in the warm season of the year, as the oil exudes more freely at that time than in cooler weather, when heavy oils and mineral tar are readily converted into a butyraceous mass. It is also best to await until water in rivers and creeks is low.

Observe whether the surface of the water exhibits variegated iridescent figures, this being especially the case in places where the water stands quietly or moves very little, as for instance, in coves. Such a film, when found, may be due to petroleum, or to iron oxides and similar substances. However, by touching the surface of the water with an iron-pointed stick, a film of oxide of iron may be disintegrated in angular pieces and small flakes, which can be moved in any direction; while oil films, when separated, re-unite, and can be readily distinguished from allied indications by the many changes in color and figures. Films of heavy oil may occasionally be seen that can be separated into angular pieces, behaving in this respect like iron oxides, but they almost invariably exhibit variegated movable rings of color. In swamps, other substances may produce a phenomenon similar to crude oil.

When indications of oil have been discovered in a quiet part of a water-course, try to remove the iridescent film, and turn up the bottom by driving the iron-pointed stick into it several times. If films of oil, together with bubbles of gas, re-appear, and this happens regularly after repeated trials, there may be a deposit of oil that should be examined further.

If these tests yield negative results, the oil must have floated down from above, and investigation of the water-course must be continued until the source of the traces of crude oil has been found. This will usually be in sandstone or other porous rock, and pieces knocked off with a hammer will reveal the oil generally in the form of drops, partly upon the surfaces of the strata and partly in small cavities. Instead of petroleum, mineral tar — a black, sticky mass — will frequently be found.

The rock itself is occasionally impregnated, which may be recognized partly by the odor and partly by the so-called ' water-test.' For this purpose place a piece of the rock in quiet water, if possible exposed to the rays of the sun; if the rock contains oil the characteristic iridescent colors appear, as a rule, upon the surface of the water.

The fresh fracture of oil-bearing sandstone is, as a rule, of a darker color than that of adjoining rock.

After rain, drops of water adhere to outcrops of oil sandstone in a manner similar to that observed on other fatty substances.

If in examining water-courses, oil-bearing sandstone be found, the problem is whether the prospector has to deal with contiguous rock or simply with an erratic block. This question can, as a rule, be decided without much difficulty, from the position of the stratification and the petrographic character of the rock when compared with the surroundings. However, if there is still a doubt, examine, by means of the water-test, the portions of rock in the natural continuation of the block.

Should the oil-bearing rock actually turn out to be an erratic block, the rock from which it has been derived will be found above, either on the slopes or in the water-course itself. Knowing the petrographic character of the oil-bearing block, it will not be difficult to find in that area the rock from which it is derived. In the foregoing manner the water-courses are traced to the limits of the region. It is advisable to examine specimens of all the sandstone by means of the water-test, since the latter frequently shows the presence of petroleum, though there may be no external indications of it.

It may be mentioned that in cooler weather, traces of oil on the surface of water do not yield blue, red, or yellow figures, or at least not very vivid ones, but a milky coloration, which possibly may also be due to other causes, so that determination is more difficult and less certain. This is another reason why it is advisable to select warm weather for prospecting. Oil may also be detected by its odor.

In swamps, iridescent films, which do not consist of iron oxides, but of hydro-carbons formed by decomposition, are occasionally noted. If due to the latter cause, they do not re-appear, or at least only to a slight extent when removed with the iron-pointed stick from the surface of the water. However, in examining the bottom, gas-bubbles generally rise to the surface. Such pools are examined first in the center, and then by detaching pieces from the edges with the stick.

Salses (mud volcanoes), as well as abundant exhalations of natural gas, if not derived from coal measures, are promising indications of the presence of petroleum.

It need scarcely be mentioned that porous rock — if oil-bearing — justifies greater expectations than compact rock; and that larger quantities of oil may be looked for in oil-bearing sandstones of greater thickness.

Although, generally speaking, a rich strike of oil may be inferred from abundant indications in the 'outcrop,' the reverse is not always correct; in many oil-fields, now productive, first indications were not especially encouraging.

If the oil occurs in definite geologic horizons, these must be sought and carefully examined in water-courses crossing them, not because the strata are most eroded there, so giving the best view of their geologic structure, but also because the oil, since the restraining covering has gone, has the best chance of exuding there. The cut of the water-course is generally one of the lowest points of the outcrop, where the most abundant exudation takes place, in consequence of the greater head or pressure.

An important question is whether the oil occurs in beds or in veins, and the following particulars may be helpful:

With proportionately greater denudation of the oil-bearing rock, it is sometimes possible to decide this question directly by observation, whereby the prospector, however, must take into consideration that even with a bed-like occurrence the oil will collect in small fissures. With a vein-like formation a fissure may be traced to where it assumes larger dimensions in the strike and dip.

If the prospector has to deal with a thick seam or stratum of sandstone, recognized as oil-bearing, imbedded in another rock — shale, for instance — such seam should be traced, and fresh pieces cut from it examined as to their oil content by the water-test. If positive results are obtained, it may be inferred that the sandstone is the bearer of the oil, and that it is a bed-like occurrence.

In a large mass of sandstone, several outcrops of oil may sometimes be found at a fair distance from one another. If in tracing the stratum of the first outcrop according to its strike, a second, third, and other outcrops are encountered, we have to deal with a bed-like occur-

rence. This tracing of the stratum is effected by means
of a compass, however, always with due consideration
to the configuration of the ground. Suppose the cross-
section of the sandstone bed with the declivity — the so-
called outcrop-line — be construed and traced. The out-

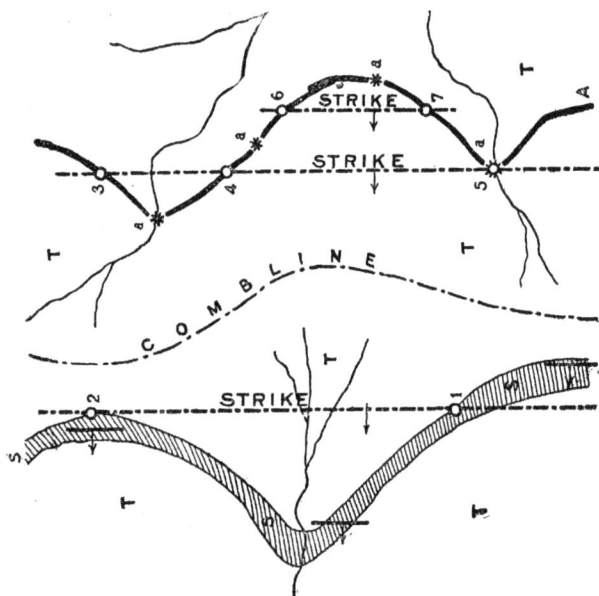

Fig. 55.

crop-line will deviate more from the straight line of strike
the flatter the strata and declivities lie. In tracing the
same stratum, it must be observed whether its strike does
not change, which, of course, will necessitate a change
in the route of the prospector.

If some promising outcrops of oil have been found,

justifying the more extensive and expensive search, it is advisable to mark accurately on a sketch-map, in addition to the outcrops, the relative heights, generally determined by an aneroid barometer, the strike and dip of the stratum reduced to the astronomical meridian, and the outcrops of well-characterized concordant strata, for instance, imbedded in shale S, Fig. 55, no matter whether they lie in the upcast or downcast of the outcrops of oil, *a*. The relative heights of one of these strata are determined in several places, selecting points that can be readily found on the map, and, if possible, lie at the same height, which can be effected without essential error with the assistance of an aneroid barometer by taking observations in rapid succession. The points of same height, 1 and 2, for instance give the strike of the stratum for a greater distance.

By connecting the outcrops of oil *a* by a line *AA,* and again determining in the latter several points of the same height — say 3, 4 and 5 — the general strike is again obtained. If the latter runs parallel with the general strike of the characteristic stratum *S,* previously traced, one is justified in inferring a bed-like occurrence of oil, even if the construed dip of the outcrop line of oil corresponds with the observed local dip of the strata.

In these investigations it is assumed that the oil is exuding from the solid rock, an error regarding the outcrop of it being, therefore, excluded. Such an error, may, however, be made when the outcrop is covered with loose earth and rock, to the base of which the oil exuding above flows down unexposed, and somehow escapes farther below.

A vein-like occurrence of oil will not show the above-mentioned conformities with the characteristic concordant strata. Such formation pre-supposes a fissure, which is generally connected with a throw of the strata. This is most frequently proved by the fact that a characteristic stratum suddenly ends; and does not re-appear in its natural level, but either to the right or left, or higher or lower. If two or more such points of disturbance have been found, their connecting line is the outcrop line of the

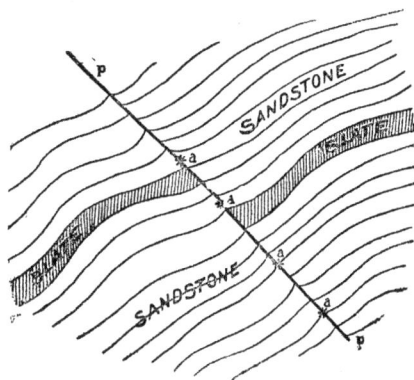

Fig. 56.

fissure, Fig. 56. If this line passes through the outcrop a, or if several outcrops lie in it, a vein-like formation of oil must be inferred.

Sometimes the oil occurs in a network of smaller and large fissures. This is shown by the fact that in the presence of several outcrops a linear distribution of the same cannot be recognized, and that the combinations yield the most varying results, according to whether exploration is carried on from the one or the other outcrop. Such occurrence presents uncommon difficulties in prospecting.

It need scarcely be mentioned that in prospecting for oil, it is of great importance to hunt up and map the anticlinals and their saddles, as well as faults.

The use of a contour map of an oil-sand to discover new pools in unprospected territory will materially aid the prospector, but cannot be absolutely relied upon, as the other conditions necessary for an accumulation can only be learned by actual test.

The directions given for prospecting may have to be modified according to local conditions. With a sufficient preliminary knowledge of geology, any difficulties will, as a rule, be readily overcome by thoroughly digesting the principles of the directions given.

Regarding the quality of the surface oil, it must be remembered that it is not a criterion for the oil at greater depth. Oil thickens on the surface of the earth, and with increasing density becomes viscous and dark. If pale, limpid, and lighter oil is found at the outcrop, it is sure evidence of oil of excellent quality at depth. In every case it may be expected that the quality of the oil at depth is superior to that at the surface.

A simple test for oil in rock or sand is as follows, according to E. G. Woodruff:

(1) Sample carefully, securing from 1 to 5 pounds. (2) Crush fine and mix. (3) Dry in the sun or by other means, but not over a fire. (4) Mix well and put a tablespoonful into a small bottle, pouring in chloroform or tetrachloride until the ore or sand is saturated and about half of the liquor is above it. (5) Cork tightly and shake for 15 minutes. (6) In a glass funnel place a white filter-paper, and a white plate below the funnel. (7) Pour the pulp onto the filter. (8) Put the dish containing the

filtrate in an open place so that it can evaporate. (9) Examine the filter, and if oil is present it will appear as a brown or black ring on the paper. (10) After the liquor has evaporated and anything remains, it is oil.

OIL-SHALE

This mineral will be a future source of oil for the United States. There are large deposits worked in Scotland; others practically idle in New South Wales, Australia; while in Colorado, Nevada, and Utah are extensive areas of the shale. Since 1917 there has been a marked revival of interest in this mineral, which had remained unworked for 80 years. Prospectors could locate claims under the placer law, but this is to be altered. In Garfield, Moffat, and Rio Blanco counties of Colorado, shale is found as practically horizontal ledges (not veins or pockets) in the upper part of Tertiary rocks known as the Green River formation. This consists of shales with narrow beds of sandstone interbedded. Rich and poor strata of shale may alternate. The former is identified from the latter by projecting along the outcrop. This is due to the oil in the shale cementing the mesas and hills, making them resistant to weathering. Some shales exude a tarry matter. Shale, unlike oil-sands, contains no free oil, but when heated, the oil is driven off and may be condensed. The average Colorado shale yields 40 gallons of oil per ton. Mining costs are fairly low. Several methods of treatment have been devised, and a few are in operation. The shales of Utah, in the Uintah region, are somewhat difficult to reach. The deposits in these States are characterized by high altitudes, precipitous and arid country.

Oil-shales, like other shales, are sedimentary in origin. They were laid down as muddy deposits in shallow seas and the remains of organic life were disseminated in them. Such deposits were originally horizontal. In the Rocky Mountain region of Utah-Colorado-Wyoming, the shales have remained almost in this position, but in many regions they have been uptilted, warped, folded, and distorted just as have been rocks of all ages and modes of origin. The grade of shale depends wholly upon the amount of organic remains — the so-called 'kerogen' disseminated through it.

Oil-shales are of several varieties, readily identified by persons familiar with such rocks. They are massive, papery, paraffin, asphaltic, sandy, and limey. Each has its peculiar texture, and behaves distinctively when subjected to heat or when undergoing pulverization. There is considerable range in the color of oil-shales, this showing from deep black through all shades of brown and blue to a light gray. The outcrop of one deposit is a light yellow, resembling ochre.

OZOKERITE

Ozokerite is a mineral paraffin or wax, and is deposited generally in fissures and cavities in the neighborhood of coal-fields and deposits of rock salt, or under sandstone pervaded by bitumen. It is not common by any means.

The most interesting deposit is in eastern Galicia, Austria. The ozokerite occurs there in a saliferous clay belonging to the Miocene of the most recent Tertiary period, and forming a narrow, almost continuous strip on the northern edge of the Carpathian Mountains. This formation consists chiefly of bluish and variegated clays, sands and sandstones, with numerous occurrences of gypsum, rock salt, and salt springs. At Boryslav, the clay strata form a perceptible saddle as they sink on the south below the so-called menilite slates, which are bituminous and foliated.

At Soldier Summit and near Colton, in Utah, is another important deposit, which has been worked.

Ozokerite has a greasy feel, is yellow-brown to green, is translucent when pure, has a light to brown streak, and specific gravity of 0.84 to 0.93. It is soluble in carbon di-sulphide. One of the principal uses is in the manufacture of gramophone records.

The odor of ozokerite is, according to its purity, agreeably wax-like. In consistence it is soft, pliable, flexible to hard; the mass in the latter case showing a conchoidal fracture, but softens on kneading. The boiling-point lies between 133 and 165° F., and of the so-called 'marble wax' even as high as 230° F.

Ozokerite is readily soluble in oil of turpentine, petroleum, benzine, etc., and with difficulty in alcohol and ether; it burns with a bright flame, generally leaving no residue. Its elementary composition is about that of petroleum, 85% of carbon and 15% of hydrogen.

Closely allied to ozokerite are the following mineral resins:

Retinite, generally a yellowish-brown, sometimes a green-yellow, or red color, is found with brown coal.

Elaterite or elastic bitumen, of a blackish-brown color, sub-translucent, and occurring in soft, flexible masses in the lead-veins of Castleton, in Derbyshire, and in the bituminous sandstone of Woodbury, Connecticut. It ranges from soft to elastic to hard and brittle. Elaterite melts in a candle flame without decripitating.

Asphalt or Bitumen is solid at ordinary temperatures. It has a black to blackish-brown color, a conchoidal fracture with glassy luster. Hardness — 1 to 2; specific gravity — 1 to 2. It melts at 90° F., and is very inflammable. It appears to be formed by the oxidation of the non-saturated hydro-carbides in petroleum. The most remarkable deposit is on the island of Trinidad, South America. It occurs also of every degree of consistence, and in immense quantity, along the coast of the Gulf of Mexico, chiefly in the States of Tamaulipas, Vera Cruz, and Tabasco, where not infrequently it is associated with rock salt and niter. It also occurs in Utah. It has been found associated with ozokerite and more extensively as melted out of sandstone. Californian shales and sandstones are characterized by bituminous matter contained therein, and layers of bitumen and seepages of tarry matter are common. In Los Angeles county is a large deposit, in which many prehistoric animals were trapped and died. Bitumen is also found exuding from the cherty limestone that contains the zinc deposits of Oklahoma. Asphalt is in great demand for paving purposes, and deposits are of considerable value.

CHAPTER XV

VARIOUS USEFUL MINERALS

Alum. This name is applied to a group of minerals which are hydrous sulphates of aluminum with potash, soda, ammonia, magnesia, etc. They all crystallize in the regular system, are soluble in water, and have an astringent sweetish taste. Hardness, 2 to 2.5; specific gravity, 1.75. Potash alum is the most common, and is usually found in the form of an efflorescence or an incrustation, with a mealy and sometimes fibrous structure. It is abundant in clays, argillaceous schists, which, when largely impregnated with alum, are called 'aluminous schists' or 'shales.'

Soda alum has a general resemblance to potash alum, but is rather more soluble in water. Magnesia alum occurs in silky-lustered fibrous masses. Iron alum forms yellowish-white silky masses. It differs somewhat from the other alums in turning red when heated.

Alum is used in dyeing and calico-printing, candle-making, dressing skins, clarifying liquids, and in pharmacy. Commercial alum is largely a manufactured product.

Apatite or phosphate of lime, with chlorine or fluorine, occurs in six-sided prisms, also in masses. It is transparent or opaque; colorless, white, yellowish, green, violet, and pink, with a greasy luster, and gives a white streak. Fracture — conchoidal or uneven; specific grav-

263

ity — 3.16 to 3.22; hardness — 5. In thin laminæ it is fusible with difficulty before the blowpipe; when moistened with sulphuric acid, it tinges the flame greenish. It is soluble in hydrochloric and nitric acids without effervescence. From beryl it is distinguished by its inferior hardness and its solubility in acids. It occurs in rocks of various kinds, but more frequently in those of a metamorphic crystalline character, as in Laurentian gneiss, which is usually hornblendic, granitic, or quartzose in character; and in association with granular limestone. It is also found as an accessory mineral in metalliferous veins, especially those of tin, and beautifully crystallized and of various colors in many eruptive rocks. It also occurs in veins by itself, mostly in limestone, but sometimes in granites and schists. In these deposits apatite is also found as concretions, sometimes showing a radiated structure, but of an earthy appearance externally.

In sedimentary formations, where a considerable accumulation of fossils has provided the phosphate of lime, apatite occurs in two principal forms, namely, coprolites, which are excreta of large animals, especially saurians, and concretions formed at the expense of the same coprolites, together with shells, bones, etc. The richest of these deposits are from Lower Cretaceous to Lower Jurassic in age, but phosphatic deposits are found and worked in sedimentary deposits of all ages.

The principal use of apatite is as a source of phosphoric acid and phosphorus, and before the discovery of phosphate-rock deposits in Florida, was largely sold to the manufacturers of fertilizers.

The Tennessee white phosphates occur in Perry and Decatur counties. The lamellar variety is the highest grade and the most easily prepared for market. It also appears to be most abundant. Selected specimens of the thin plates contain 85 to 90% of lime phosphate. The less dense greenish material, which is associated with the white and pink plates, contains some ferrous iron, and runs slightly under 80% of lime phosphate. In Amelia county, Virginia, apatite is found with mica and beryl in mica mines; and in Morris county, New Jersey, mixed with magnetite.

Arsenic is found in the mineral kingdom partly in a metallic state (rare), partly in combination with oxygen, sulphur, and other elements.

Native arsenic occurs seldom distinctly crystallized, but usually in fine granular, spherical or nodular masses. Its specific gravity is 5.63 to 5.73; hardness, 3.5; brittle; uneven and fine-grained fracture; metallic luster; color and streak are tin-white, usually with a grayish-black tarnish; evolves an odor of garlic on breaking; contains occasionally more or less iron, cobalt, nickel, antimony, and silver.

Before the blowpipe it volatilizes quickly before fusing, giving off white fume having an odor of garlic. Native arsenic occurs especially in veins in crystalline slates and transition rocks in subordinate quantities associated with ores of gold, silver, lead, cobalt, and nickel.

Realgar, with 70.029% of arsenic and 39.971% sulphur, has a red color; crystallizes monoclinic; fracture, conchoidal to splintery; hardness, 1.5 to 2.0; and specific gravity, 3.4 to 3.6. It is but slightly affected by acids; soluble with a deposit of sulphur in aqua regia, and in concentrated potash lye with separation of dark brown sulphide of arsenic. From ruby silver and cinnabar, it is readily distinguished by its inferior hardness, lower specific gravity, and orange-yellow streak, the streak of the two above-mentioned minerals being cochineal-red.

Orpiment, with 69.9% of arsenic and 39.1% of sulphur; exists in nature, but for industrial purposes is mostly artificially prepared. The mineral has a lustrous lemon-yellow or orange-yellow color, is cleavable into thin, flexible, transparent laminæ; hardness, 1.5 to 2; specific gravity, 3.4 to 3.5; soluble in nitric acid, potash lye and ammonia.

Arseno-pyrite, or mispickel, is a sulph-arsenide of iron, contains 46% of arsenic. It is a common vein mineral and is sometimes highly auriferous, especially in California. It is common in crystals, has a silver-white to steel-gray color; grayish-black streak; metallic luster; hardness of 5.5 to 6; and specific gravity of 5.9 to 6.2. Most of the white arsenic of commerce is obtained from this mineral as a by-product condensed in the flues of smelters reducing copper, gold and other ores.

Asbestos. Color, green or white. The asbestos of commerce is practically a finely fibrous form of serpentine or amphibole, that is to say, it is essentially a hydrated magnesium silicate. Every deposit of serpentine is a possible repository of asbestos. It occurs in seams half an inch to several inches in width, running parallel to or crossing one another, the width of each seam making the length of the fiber. Canada furnishes most of the world's supply of asbestos, mining being confined to a small area in the great serpentine belt of the Province of Quebec, that lies to the south of the St. Lawrence river. Near Globe, Arizona, some excellent asbestos has been mined, and a little in Trinity county, California. As a slow conductor of heat and an incombustible material asbestos has many uses.

Barite or barytes, sulphate of barium. This is commonly called 'heavy spar,' occurs in tabular, glassy crystals, and also in dull masses in veins of various rock formations. Color, white or tinted; transparent or translucent; luster, vitreous or pearly. Its specific gravity is 4.3 to 4.7; and hardness, 2.5 to 3.5. It is readily distinguished by its great comparative weight. When heated in a blowpipe flame splinters fly off the crystals. It fuses with difficulty, and imparts a green tinge to the flame. After fusion with soda, it stains a silver coin black. It is not acted upon by acids.

Barite and scheelite resemble one another, so to distinguish them put a piece of each into a fire, when the former flies to pieces and the latter does not. Barite contains sulphur, and if a little be mixed with soda, moistened with water, and heated on charcoal by blowpipe, the melted mass when put on a clean silver coin will blacken it.

In the United States, barite is found in many places, it being mined in Alaska, Virginia, Missouri, New Jersey, and other States. It frequently occurs in connection with lead and zinc deposits, forming the gangue of the metal-bearing vein. The best varieties of barite are the white and gray. The chief use of barite is as a pigment, as a cheap substitute for white lead. It is also used for weighting paper and leather.

The carbonate of barium, witherite, is a much less common mineral than the sulphate. It sometimes occurs in crystals, but the more common form is that in fibrous masses. It exists in veins. It fuses easily in the forceps, and gives a yellow-green flame. In hydrochloric acid, barite dissolves with effervescence, the solution yielding a heavy white precipitate (barium sulphate) if a little sulphuric acid be added. Witherite is used in the refining of sugar, and also in the manufacture of plate glass.

Borax. This is a hydrous borate of sodium, occurring as crystals, powder and incrustations. Its hardness is 2 to 2.5; specific gravity, 1.7; fracture, conchoidal; luster, vitreous to resinous; color, white, sometimes grayish, bluish, or greenish; streak, white; and taste, slightly alkaline and sweetish. It is translucent to opaque. The occurrence of deposits of borax in the United States is limited to California, Nevada, Oregon, and Texas, but the first State is the principal source of the world's supply. Chile produces a fair quantity. In the early seventies, borax in large quantity and in a pure condition was discovered on many of the alkaline marshes of western Nevada and eastern California. Refining plants were established and flourished for a time.

In 1890 it was found that the borax crust on most of the marshes is a secondary deposit, being derived from the leaching of beds of borate of lime in the Tertiary lake sediments, which abound in the region. The marshes were abandoned and a mine was established on a bedded deposit at Borate near Daggett, San Bernardino county, California. This has since been exhausted. The value of this deposit led to the discovery in Death Valley * of

* Death Valley received its sinister name from the fact that, in 1849, a band of emigrants wandered into the valley, and most of them perished from thirst before an avenue of escape was discovered.

large quantities that far exceed those worked near Daggett.

The borax of Death Valley, as well as that near Daggett, occurs in a regular stratum, inter-bedded with the semi-inducted sands and clays that make up the bulk of the strata. These beds are generally regarded as of Tertiary age, and they are supposed to have been deposited in enclosed bodies of water.

The principal deposits of boron salts are on Furnace creek, and at Resting Springs, from where it was hauled to Mojave by mule-teams. The mineral found here is borate of lime, or ' colemanite,' and it occurs as a bedded deposit from 5 to 30 feet in thickness, inter-stratified in lake sediments. These lake beds are composed of semi-indurated clays, sandstones and coarse conglomerates with inter-calated sheets of volcanic tuff and lava. The rocks are severely folded, the axes of the folds lying in an east-west direction.

Colemanite. This is a hydrous borate of calcium, and is the main source of borax in California. It occurs as crystals and massive, with a white to yellowish-white color. Luster is vitreous; hardness, 4 to 4.5; and specific gravity, 2.42.

A considerable quantity of borax is now obtained from the brine of Searles Lake in San Bernardino county. This brine is treated for potash; the quantity available is enormous.

Borax is used in medicine and as an antiseptic by meat packers and others. Its chief use, however, is as a flux in metallurgical operations, in enameling, glazing of pottery, and in the manufacture of glass.

Borax may be detected by dissolving a little of the suspected material with a small quantity of water, add a few drops of sulphuric acid, then some alcohol (if procurable) in some form, and lighting the mass. Borax is shown by a bright-green flame. Search for this mineral is always worth-while, as the demand is continuous.

Clays. These are all products of alteration from other minerals. Their composition is variable, and they do not crystallize. The true clays are all plastic and refractory to a greater or less degree, and on these properties depends their value for industrial purposes. Pure kaolin is the best of all clays.

The presence of alkalies in clays is objectionable, as it renders them fusible, as also do many other oxides. Iron is not only objectionable on the score of fusibility, but also as coloring matter. The presence of too large a proportion of water, carbonic acid gas, or organic matter, causes clay to contract under the action of fire, and the same result will follow if the clay is partly fusible.

The soft clays are divided as follows:

Kaolin, porcelain clay or china clay. This is a product of decomposition of feldspar and other minerals, and never occurs in any crystalline form. Its composition varies somewhat according to the source from which it has been derived. In all cases it is a hydrated silicate of alumina, and its usual source is feldspar. It is a friable, soft substance of a white yellow, or flesh-red color, and capable of resisting the highest heat of a porcelain furnace. It usually contains more or less silica in an uncombined state. Its specific gravity is 2.6, and hardness, 2 to 2.5. Kaolin is almost entirely from the older feldspathic rocks, while clays are generally derived from younger rocks.

Pottery or plastic clay is not so pure as kaolin, containing a large amount of iron.

Bole is a hydrated silicate of alumina and iron, of a somewhat variable composition, but generally containing about 42% of silica and 24% of water. It also contains a large amount of ferric oxide, which gives it its yellow-red or brownish-black color. It is soft and greasy, translucent or opaque, adheres to the tongue, and falls to pieces with a crackling noise when immersed in water. The hardness is 1.5, and specific gravity, 1.4 to 2. It fuses with facility into a greenish enamel.

Fuller's earth. This is a kind of clay composed, when pure, of 45% silica, 20 to 25% alumina, magnesia and water. It was formerly largely used as an absorbent in fulling or freeing woolen fabrics and cloth from fatty matters, but other substances have been substituted for this purpose. Its main use now is in filtering and decolorizing oils and fats.

Extensive deposits of fuller's earth occur in Decatur county, Georgia; and in Gadsden, Leon and Alachua counties, Florida; and in Arkansas and Texas.

Coal. This is a mineral and normally the world produces one billion tons a year. It is massive and uncrystalline. Color is black or brown; and is opaque. It is brittle or imperfectly sectile. Hardness is 0.5 to 2.5; and specific gravity, 1.2 to 1.80. Coal is composed of carbon with some oxygen and hydrogen, more or less moisture, and traces also of nitrogen, besides some earthy material, which constitutes the ash.

The greater part of coal formations originally consisted of peat or accumulations of vegetal matter in varying stages of preservation, with small proportions of the remains of animals. Sand and clay were collected in the deposit, and sometimes hydrated oxide of iron or ferrous carbonate. These deposits, in the course of time, with pressure, were formed into strata of alternate sandstone (usually gray) and slate-clay or shale; and between these strata, beds of brown or black coal or anthracite and clay-ironstone were formed, subordinate, however, in extent and thickness to the sandstone, slate, and shale. Coarse conglomerates, marl, or limestone rarely occur in these formations. The Carboniferous and the Tertiary periods furnish the most characteristic examples of these formations; but the carbonaceous deposits of other periods are associated with similar rocks, and are so like the genuine coal formations that, petrographically, they are hardly to be distinguished from them.

Anthracite (glance coal, stone coal). Its luster is high, not resinous, sometimes sub-metallic; color — gray-black; hardness — 2 to 2.5; specific gravity, if pure — 1.57 to 1.67; and fracture is often conchoidal. Anthracite consists almost entirely of carbon, and contains little hydrogen, oxygen, or nitrogen. It contains earthy admixtures in various quantity. The normal position of anthracite appears to be in the transition formations. Enormous anthracite seams or beds are worked in Pennsylvania and in Wales, England.

Bituminous coal. Color — black; luster — usually somewhat resinous; hardness — 1.5 to 2; specific gravity — 1.2 to 1.4; and contains usually from 75 to 85% of carbon.

Cannel coal. Is very compact and even in texture, with little luster, and fracture largely conchoidal.

Brown coal (often called **lignite**). Color, black to

brownish-black. Contains 52 to 65% of fixed carbon. In Europe such coal is briquetted before use.

Peat. This is not a mineral, but consists of the cumulatively resolved fibrous parts of certain mosses and graminaceæ. It gradually darkens from brown to black with increasing age. It occurs in beds or in bogs. As a fuel it is most economically used at the place where it is grown. Good peat yields about 3 to 6% of tar, which is comparatively easy to purify by the usual method.

Examination of a peat bog is very instructive in the formation of coal as affording examples of vegetal matter in every stage of decomposition; from that in which the organized structure is still clearly visible, to the black carbonaceous mass that only requires consolidation by pressure in order to resemble a true coal.

There are large deposits of peat in Canada, Europe, and the United States, especially in North Dakota. Investigations during recent years show that after being dried, peat is a good fuel, yields gas for engines, is a fertilizer and food for stock, is suitable for surgical dressings and packing goods.

Jet resembles cannel coal, but is harder, of a deeper black and higher luster. It takes a brilliant polish and is set in jewelry.

Dolomite is composed of carbon dioxide, lime, magnesia. It occurs in rhombohedrons, faces often curved. It is frequently granular or massive; white or dull tinted; and glassy or pearly. The specific gravity is 2.8 to 2.9; and hardness, 3.5 to 4. It effervesces in nitric acid, and dissolves more slowly than calc-spar. Dolomite yields quicklime when burned or calcined. It occurs in extensive beds of various ages like limestone, principally the serpentine. It is used as a building-stone and in the manufacture of Epsom salts. It is difficult to distinguish from calcite without chemical analysis.

Feldspar. This is composed of silica, 64.20%; alumina, 18.40%; and potash or soda (lime), 16.95%. It is crystallized or in irregular masses; opaque; usually flesh-red or white, or of various dull tints; has a glassy or pearly luster; irregular fracture, but in some directions it splits with an even, glimmering cleavage face; specific gravity of 2.3 to 2.8; and hardness, 6. Before the blowpipe it fuses with difficulty, and is not touched by acids. Where found in sufficient quantity to be of industrial value, it is usually obtained from veins in granite or pegmatite. It exists in nearly all igneous rocks. The minerals associated with feldspar are chiefly quartz and mica; while tourmaline and topaz are common components. Feldspar is, to a limited extent, employed in the manufacture of glass, but the chief use for it is as a china glaze and as a glass-forming ingredient in the body of porcelains. As a source of potash it has possibilities.

There are a number of feldspars, all rock-forming minerals, which are silicates of alumina with potash, soda, and lime, having similar characteristics. The names are orthoclase, microcline, anorthoclase, albite, oligoclase, andesine, labradorite, bytownite, and anorthite.

One of the finest varieties of feldspar is that known as 'adularia,' from Mt. Adula, near the St. Gothard Pass, where it is found re-deposited from the rock mass in veins and cavities. It consists of silica, 64%; alumina, 20%; lime, 2%; and potash, 14%. 'Moonstone' is another variety, with bluish-white spots of a pearly luster. 'Sunstone' is another, with a pale yellow color with minute scales of mica. 'Aventurine,' feldspar sprinkled with iridescent spots from the presence of minute particles of titanium or iron. The last three varieties are employed as gem-stones, being occasionally set in brooches, but are too soft for rings.

A beautiful variety of orthoclase known as 'amazon stone' occurs in large green crystals near Pike's Peak, in Colorado, in Siberia and elsewhere.

Flint consists of silica, which in a fine condition has been separated from the surrounding rock, and which, attracted to some organic or inorganic nucleus, and sometimes only to itself, has grown in successive layers or bands, often of different colors. Flint breaks with a conchoidal fracture and cutting edges. 'Chert' and 'chalcedony' are allied to flint, but are more brittle and it takes its color — dirty gray, red, and reddish-yellow, green or brown — from the rocks in which they are found. They occur in portions of sandstone rocks usually containing a little lime, the fine silica being seemingly collected into one spot. In the chalk cliffs of Dover, England, and along the Baltic are large quantities of smooth flint pebbles. These are gathered, and used in tube-mills at gold and silver mines in all parts of the world.

Fluorspar, fluorite, consists of 48.7% of fluorine and 51.3% of calcium. Its usual mode of occurrence is massive and granular. It is also found as cubic crystals in vugs or coating the walls of small fractures in the country rock. It is generally translucent, though rarely transparent; its color is white or light violet, blue, purple, and occasionally yellow; sometimes layers of different tints in the same piece. Luster is glassy. Fluorspar breaks with smooth cleavage-planes parallel to the octahedral faces.

The specific gravity is 3 to 3.2; and hardness, 4. Before the blowpipe it is fusible with difficulty to an enamel. Fluorspar is used in the manufacture of hydrofluoric acid, with which glass is etched, in hydrofluosilicic acid, as used at Trail, British Columbia, as a flux, and in enamel- ing. Sometimes it is employed for ornaments, especially massive pieces, they taking a high polish.

At Wagon Wheel Gap, Mineral county, Colorado, the vein in which the extensive bodies of fluorspar occur, is a fissure, or fissured zone, in rhyolite tuffs and breccias, and possibly in some places in the solid rhyolite flows. The wall-rock is highly altered. The vein-filling consists of fluorite, barite, altered country rock, and gouge. The fluorite is white and massive, with a sugary appearance; or is translucent, colorless to deep purple, crystalline crusts.

In Illinois and Kentucky, fluorspar is associated with galena, two marketable products easily separated from one another.

Graphite, plumbago, black lead, consists essentially of carbon, in mechanical admixture with varying proportions of silicious matter, as clay, sand, or limestone. It occurs in hexagonal crystals, but usually in foliated or massive layers. Color, steel-gray to bluish-black. Hardness — very slight, from 0.5 to 1. Soils the fingers, makes a mark upon paper, and feels greasy. The specific gravities of different kinds of graphite vary according to the content of foreign admixtures, but lie within the limits of 2.09 and 2.23. Graphite is not affected by acids and strongly resists other chemical agents. It is largely used in the manufacture of pencils, crucibles, foundry molds, stove polish, and lubricants for heavy machinery.

Geology. Graphite is found in various parts of the world, chiefly in crystalline limestone, in gneiss and mica-schists, frequently replacing the mica in the latter so that they become actual graphite-schists. It is also found as layers in some schists and gneisses, and when in quantity these are mined for the graphite. In gold-quartz veins — such as the Mother Lode in California, West Africa, and Kalgoorlie, Western Australia — the graphite is a nuisance, interfering with milling. Two distinct varieties are noticed: the one fine-grained, or amorphous; the other foliated, or compounded of numerous little scales. Sometimes it appears as an impregnation of other rocks rather than as a distinct rock in itself. Geologically, it is confined to the oldest formations, and is usually, if not universally, associated with metamorphic action. The chief source of the mineral for many years was in Ceylon. The soil and rocks of that island are almost everywhere impregnated with graphite, so that it may be seen covering the surface after a shower. The supply is very large, the mineral existing in such quantities in the gneissic rocks that, upon their decomposition, it is seen in bright silver-like specks throughout. Ceylon graphite is particularly remarkable for its purity, containing as it does very small proportions of silicious ash.

In recent years the island of Madagascar has been producing large quantities of high-grade graphite.

Graphitiferous rocks of the Laurentian system are widely spread throughout Canada and some parts of the United States. The graphite in these rocks usually occurs in beds and seams varying in thickness from a few inches to 2 or 3 feet. Perhaps the most important and extensive of the Canadian deposits is that near the township of Buckingham, Quebec, where the graphite occurs both in beds or veins, or is disseminated.

Considerable deposits of graphite are found in Chester county, Pennsylvania; Essex county, New York; in Alabama, and Texas. American deposits received much attention from the Government during 1917 and 1918. There are several large graphite producers in these States.

Close to Ticonderoga, Essex county, New York, graphite is obtained from a hill, locally known as Black-lead mountain. The graphite beds are inter-stratified between gneissic rocks. The beds dip at an angle of 45°. The ore in them is chiefly of the foliated variety, and is mixed with gneiss and quartz in the beds in veins or layers from 1 to 8 inches in thickness, some of the deposits being richer than others. One of these has been followed to a depth of 350 feet. It is found of varying thickness, and opens out at times into pockets.

Graphite is said to occur in great purity in different localities in Albany county, Wyoming, in veins from 1½ to 5 feet thick. At Pitkin, Gunnison county, Colorado, it occurs massive in beds 2 feet thick, but of impure quality. It is also found in the coal measures of New Mexico, in Nevada, in Utah, at Dillon, Montana, and in the Black Hills of South Dakota.

The value of graphite depends upon the percentage of its carbon. To test the purity of graphite, pulverize 20 grains and then dry at about 350° F.; then place it in a tube of hard glass 4 to 5 inches long, half an inch wide, and closed on one end. Add 20 times as much dried oxide of lead and mix intimately. Weigh the tube and contents, and afterwards heat before the blowpipe until the contents are completely fused and no longer evolve gases. Ten minutes will suffice for this. Allow the tube to cool, and weigh it. The loss in weight is carbonic acid gas. For every 28 parts of loss there must have been 12 of carbon.

Gypsum is a hydrous sulphate of lime, and is composed of sulphurous acid, lime (32.5%), and water. It occurs in prisms with oblique terminations, sometimes resembling an arrow-head. It is transparent, or opaque, white, light-brown, or reddish, with a glassy, pearly, or satin luster. Cleavage occurs easily in one direction; specific gravity, 2.3; hardness, 1.5; and can be cut readily with a knife. In the blowpipe flame it becomes white and opaque without fusing, and can then be easily crumbled between the fingers. Nitric acid does not cause effervescence. It occurs in fissures and in stratified rocks, often forming extensive beds; also on beds of desert lakes. The evaporation of lime sulphate waters forms gypsum. These last two occurrences are likely to contain impure mineral. When pure white it is called alabaster; when transparent, selenite; and when fibrous, satin spar. When burnt, gypsum loses its water and falls to powder. This powder, called plaster of Paris, which is perfectly white when free from iron, possesses the property of re-absorbing the water lost, and in a very short time of assuming again the solid state, expanding slightly in so doing. It is this last property that renders the plaster so valuable for obtaining casts. It is also used raw and ground as a fertilizer, from 200 to 500 pounds being sprinkled on an acre of land.

Infusorial earth is an earthy, sometimes chalk-like silicious material, entirely or largely made up of the microscopic shells of the minute organisms called ' diatoms.' It occurs in beds sometimes of great extent, sometimes beneath peat beds, and is obtained for commerce in Maine, New Hampshire, Massachusetts, Virginia, California, Nevada, Missouri. In the last named State, farmers mine the tripoli, as it is called there, during the off season. It feels harsh between the fingers and is of a white or grayish color, but often discolored by various impurities. Infusorial earth is used as a polishing powder, ' electro-silicon ' being the trade-name of one kind much used for polishing silver. It is also used for making soda silicate and for certain cement. Being a poor conductor of heat, it is applied as a protection to steam boilers and pipes. It is also employed for filling soap, as a filter for water, and was once consumed as an absorbent for nitroglycerine in explosives, when it was commonly called diatomaceous earth.

Limestone. This is a very plentiful rock, used in enormous quantities for portland cement, plaster, fertilizer, blast-furnace flux, a source of carbonic acid gas, and many other purposes. Limestone is a carbonate of calcium, containing 56% calcium oxide and 44% carbon dioxide. Crystal forms include rhombohedrons, and it is massive, granular, stalactitic, and granular. Calcite has many colors and has a vitreous luster. Its hardness is 3, and specific gravity, 2.71. Some of the important forms of limestone are marble, onyx marble, travertine, and chalk. Chalk beds should always be examined for flint pebbles or boulders, as in Dover, England, and along the Baltic Sea in Europe. These are always in demand by mines that use tube-mills. A few drops of muriatic acid on limestone causes an effervescence. A calcareous marl (shell), a variety of calcite, is being increasingly used as a fertilizer.

Lithographic limestone is the only stone yet found possessing the necessary qualifications for lithographic work. It is a fine-grained homogeneous limestone, breaking with an imperfect shell-like or conchoidal fracture, and, as a rule, of a gray, drab or yellowish color. A good stone must be sufficiently porous to absorb the greasy compound that holds the ink, soft enough to work readily under the engraver's tool, yet not too soft, and must be firm in texture throughout and entirely free from all veins and inequalities. The best stone, and indeed the only one which has yet been found to fill satisfactorily all these requirements, occurs at Solenhofen, Bavaria. These beds are of Upper Jurassic age, and form a mass 80 feet thick. The prevailing tints of the stone are yellowish or drab. In the United States lithographic stone is found in Alabama, Iowa, Kentucky, Nebraska, South Dakota, and Tennessee, but the only deposit of importance is at Bradensburg, Kentucky.

Magnesite. When prospecting is being done in an area of serpentine (silicate of magnesia) rocks, besides seeking for chromite a good look-out should be kept for magnesite. This is a non-metallic mineral much in demand for refractory purposes, plaster, a filler in paper, paint, and chemical purposes. Before being used it is calcined or 'burned' to rid the ore of the carbon dioxide, which constitutes 52.4%, the remainder being magnesia — really a carbonate of magnesia. Crystals are rare; color is white to brown; fracture is concoidal; hardness is 3.5 to 4.5; and specific gravity is 3 to 3.12.

Dolomite, the carbonate of magnesium and calcium, is

often mistaken for magnesite; so is calcite, the carbonate of lime. The hardness of all three is somewhat similar, but their specific gravity is lower than magnesite. A drop of hydrochloric acid effervesces on dolomite and calcite, but not on magnesite.

Most of the American magnesite deposits are in California, where they are found as veins or lenses of varied size in massive serpentine. The outcrops are white and may be seen from considerable distance. Many veins have well-defined walls. Two or three deposits carry 2 to 4% iron, making the ore more desirable for refractory purposes. Over 3% silica is undesirable. The largest deposits in this country are in Washington, near Chewelah, north of Spokane. There the magnesite is crystalline and occurs as extensive beds in a sedimentary series in which are found dolomite, shale, and quartzite, into which basic igneous rocks have been intruded. Austria and Greece used to supply a desirable product, but domestic deposits and quality now seem to be ample.

Meerschaum or Sepiolite is a magnesium silicate. When pure, it is very light; and, when dry, it will float upon water. It will be recognized by its property, when, dry, of adhering to the tongue, and by its smooth, compact texture. Its hardness is 2 to 2.5, and specific gravity, 2. It is generally found in serpentine, in which rock it occurs in nodular masses but it is also found in limestones of Tertiary age. It is of snow-white color and a useful substance when found in quantity, being much employed for the bowls of tobacco pipes, and for this purpose is mined in Asia Minor.

Micas. These are silicates of alumina with potash, rarely soda or lithia, also magnesia, iron, and some other elements. They are important rock-forming minerals of igneous and metamorphic rocks. Always crystallized in thin plates, which may be split into extremely thin, flexible layers. Transparent in thin layers. The color is white, green, brown to black; specific gravity, 2.7 to 3.1; hardness, 2 to 2.5; and is easily scratched with a knife. Before the blowpipe mica whitens, but is infusible except on thin edges. When it can be obtained in large sheets, it is valuable. It is sometimes used in place of window-glass on board ship, for stoves and for lamp-chimneys. The ground material is used as a lubricant and in making ornamental and fire-proof paint. The largest consumption of mica is in the electrical trade, where it is used as an insulator on generators, motors, and at contacts.

Biotite, or black mica, contains more magnesia than alumina. It is often present in eruptive rocks, especially granites and schists. It is black to green in color. 'Muscovite,' or white mica, on the contrary, contains more alumina than magnesia, and as it also contains potash in small, but appreciable quantities, it is sometimes called 'potash mica,' and biotite 'magnesian mica.' Muscovite is an important mineral to the tin miner, since it is always found in that class of granite in which cassiterite exists, and with quartz alone forms the rock called 'greisen,' which is generally associated with tin. The rock in which large sheets of mica are found is called by some geologists 'pegmatite,' and has the same composition as granite itself, but the crystals are of a larger size.

Quebec contributes most of the world's mica. The United States has a number of deposits but not of great importance.

Monazite. This valuable mineral has been found in black sands of rivers in America, but the principal source is Brazil. It is a phosphate of cerium, didymium, and lanthanum, carrying the rare earth thorium. The value of monazite lies in the thorium and cerium, which are the salts used in impregnating incandescent gas mantles. Recently, another use has been derived from monazite, that in extracting mesothorium, which to a certain extent, may be used for some of the purposes wherein radium is necessary.

Niter. There are two forms of this valuable salt, which is used largely in explosives, namely, potassium nitrate and sodium nitrate. The former was mined mostly in India, and is white, has a vitreous luster, salty taste, hardness of 2, specific gravity of 2.1, and occurs as incrustations; while the latter is white, reddish, and yellowish, occurs as massive crystals and incrustations, has a vitreous luster, hardness of 1.5, and specific gravity of 2.24. The sodium salt is known as Chile saltpeter, and normally is exported from that country at the rate of 40,000,000 quintals (101 pounds each) per annum. Nitrates can only exist in solid form in arid regions. In Oregon they have been detected as incrustations on rocks in the east-central part, but are not considered of commercial value. Both nitrates are soluble in water. The potassium salt fuses easily on platinum wire before the blowpipe, tinging the flame green; while the sodium salt tinges it yellow. The latter deflagrates in glowing charcoal, but less vividly than the other salt.

Phosphate rock. There are large quantities of this useful mineral in America, especially in Florida, Tennessee, Idaho, Utah, and Wyoming. In the first named it occurs as land pebble and hard rock, the annual output being over 2,000,000 tons. In Utah and Idaho it is in carboniferous limestone. Phosphate rock or lime phosphate is used as a fertilizer, and to make it soluble it is treated with sulphuric acid, which converts it into a tri-calcium phosphate. As much as 250,000 tons of acid is used annually by one firm in the Southern States.

Search for phosphate rock in the Rocky Mountains should be confined to rocks, generally limestone, near the top of the carboniferous system. A field test consists of crushing the rock to pass a 100-mesh sieve. As much powder as will cover a 25-cent piece is put in a small enameled cup, and 30 cubic centimeters of water and 10 of concentrated nitric acid added. Filter, or decant the fluid if it is clear, into a glass beaker, and add 100 c.c. of water, then a little saturated solution of ammonium carbonate. This will probably make the clear solution somewhat cloudy. Nitric acid should then be added drop by drop until the solution clears again, and gives a faint but distinct acid reaction with blue litmus paper. The solution is then warmed to a temperature of 70 or 80° centigrade (158 to 176° Fahrenheit), and 50 c.c. of a concentrated solution of ammonium molybdate is added, drop by drop, while the solution is being stirred. This solution is allowed to stand in a warm place for 15 minutes; if phosphoric acid is present a bright yellow granular precipitate of phosphor-molybdate of ammonia will appear. The above manipulation is not difficult.

Potash. This is mostly used as a fertilizer. Prior to 1914, Germany supplied the bulk of it, and since that time by strenuous efforts the United States can produce 20% of its consumption. In 1913, over 1,000,000 tons was im-

ported.   In Europe, potash rock is mined at great depths.
In America the principal sources are brines from alkali
lakes in California and Nebraska, kelp or seaweed, alunite
(a hydrous sulphate of aluminum and potassium, containing 11.4% potash, a pink crystalline rock found in
quartzite, particularly in Utah), and leucite (a silicate
rock) in Wyoming.   Lands containing potash were
reserved by the Federal Government until September,
1919, when Congress passed a bill authorizing the
leasing of such areas, also those containing phosphate
rocks.

Rock Salt or halite has the character of ordinary table
salt, but is more or less impure.   Occurs in beds inter-
stratified with sandstone and clays, which are usually of
a red color and associated with gypsum.   The specific
gravity is 2.5 and hardness 2.13.   It contains 39.30% of
sodium and 60.66% of chlorine, but most samples contain
clay and a little lime and magnesia.   The surface indications or rock salt are brine springs supporting a vegetation like that near the sea-coast, also occasional sinking of
the soil caused by the removal of the subterranean bed of
salt by spring water.   Rock salt is obtained by sinking
wells from which the brine is pumped and evaporated in
large pans, or by mining, the same as for any other ore.
Salt deposits occur in the strata of all ages, from the
Silurian onward.

Most of the salt in North America is obtained from
brine wells in Michigan.   Valuable and productive springs
are worked in the Syracuse and Salina districts, New
York, and in Ohio.   Some of these arise from a red
sandstone whose geologic horizon is said to be below the
coal measures.   Lakes in arid regions, such as in Inyo

county, California, and Great Salt Lake, Utah, yield a considerable quantity of salt by the solar process — evaporation. So do the salt beds in China, where the salt industry is a government monopoly.

The famous salt mine of Wieliezka, near Cracow, in Galicia, has been worked since the year 1251, and it has still enormous reserves of the mineral.

Slate is an argillaceous shale easily recognized by its parallel cleavage, and varies in color from light sea-green and gray to red, purple, and black. It has been formed by sedimentary deposits, and now constitutes extensive beds in the Silurian formation. Good deposits of slate are always in demand.

Sulphur and Pyrite. Native sulphur or brimstone occurs crystallized or massive in volcanic regions and in beds of gypsum. It is frequently found contaminated with clay or pitch. The principal uses of sulphur are in making sulphite for the paper industry and for sulphuric acid. Color, yellow; luster, resinous; specific gravity, 2.1; hardness, 1.5 to 2.5. It is fusible, burns with a blue flame, and gives off a well-known odor. The island of Sicily used to supply most of the world's demand for sulphur, but in recent years Japan has contributed to the supply, and America produces all it requires. The most important deposits of sulphur in the United States are in Louisiana and Texas, where the mineral occurs associated with limestone above gypsum at a depth of several hundred feet. Boreholes or wells are sunk, and the sulphur melted out of the rock by superheated water, known as the Frasch process. The product is pure sulphur, and the output is large.

Sulphur is found associated with rhyolite in Nevada,

and in travertine in Wyoming; also in Beaver county, Utah.

A scarcity of sulphur has led to greater attention being paid to pyrite and pyrrhotite, especially for the manufacture of sulphuric acid. While there are many deposits of iron pyrite in most parts of the world, they are not always accessible for mining at a low cost, and situated so that transportation of the low-valued product is easy and cheap. These primary conditions are essential to the industrial utilization of any pyrite bed.

An enormous quantity of sulphuric acid is now made from the fumes from blast and reverberatory furnaces reducing copper, lead, and zinc ores. One plant in Tennessee makes 1000 tons of acid daily by using the fume.

Pyrite. This is a common brass-yellow mineral, found in all classes of rock. It is prominent in schist, sandstone, slate, and quartzite. Crystals are common, large, and are usually in cubes. In gold areas large crystals rarely carry much precious metal. Gold in pyrite is mechanically, not chemically, mixed. The ' iron hat ' or gossan copper deposits is due to alteration of the pyrite. Large deposits of pyrite are nearly always associated with copper. It has a metallic luster, greenish-black streak, specific gravity of 5, and hardness of 6.5. Pyrite is a sulphide of iron. For commercial purposes it should carry at least 30% sulphur.

Pyrite cannot be scratched with a knife, but is scratched by quartz, and scratches glass with great facility. Before the blowpipe it burns with a blue flame, giving off an odor of sulphur, and ultimately fuses into a black magnetic globule. It is easily distinguished from copper pyrite by its hardness, the latter being readily cut with a knife. From gold it is distinguished by its hardness and in not being malleable, and in giving off sulphurous odors in the blowpipe flame.

Arsenical pyrite or mispickel contains 34.4% of iron, 19.6% of arsenic, and 46.0% of sulphur. It occurs in flattened prisms and also massive. Its color is white; luster, metallic; streak, gray; fracture, uneven; specific gravity, 6 to 6.3; hardness, 5.5; and cannot be scratched with a knife, but is scratched by quartz. Heated before the blowpipe, it gives off white arsenical fumes of a garlic odor, and finally fuses into a black globule. It is abundant in mining districts, and sometimes is auriferous. With the improved processes now in use, it is possible to extract the gold profitably, and hence mispickel ores should be examined for gold. Mispickel is common in Californian ores; also at Manhattan, Nevada; and in Western Australia.

In Wisconsin, pyrite is abundant in some of the lead and zinc mines, where it is associated with galena and sphalerite. In California there are veins of pure pyrite and deposits in the Shasta copper belt. Leadville, Colorado, produces the mineral. In St. Lawrence county, New York, are large quantities as lenses or veins in gneiss and schist, averaging 21% sulphur. Near Sherbrooke in Quebec, Canada, are four parallel interbedded lenses of cupriferous pyrite, in talcose-schist crossed by diorite dikes. While this ore carries 2½% copper, it is primarily used for the 40% sulphur contained. There are also large deposits in quartz in Georgia. In many States pyrite is found in gold and silver districts, where it is rarely enough to be worked for its sulphur content.

The old Haile gold mine in South Carolina, opened in

1830, is now a pyrite producer. The old volcanic tuffs and porphyry have been changed into alternating bands of almost pure sericite and highly silicious quartz-sericitic schists. The whole series of schists is more or less impregnated with pyrite, and along some zones this mineral is highly concentrated. The average sulphur content of the pyrite is 25%. Where the pyrite content of the ore is high, especially in soft schists, gold is correspondingly low.

Near Oakland, California, is a body of pyrite of good grade lying between serpentine and altered volcanic rock. This is mined and converted into sulphuric acid. Large quantities of pyrite, called coal brasses are found in coal mines, and have generally been discarded as waste. Some mines occasionally gather a carload of hand-cleaned pyrite, but few bother with it. The mineral when cleaned carries enough sulphur, but too much carbon, an objectionable feature.

At Foldal, Norway, a cupriferous pyrite is extracted and concentrated, yielding a 44% sulphur product; while the largest deposit of pyrite in the world is in the Rio Tinto district of Spain, where the copper-bearing mineral is disseminated in a silicified porphyry gangue.

Pyrrhotite. This is often found with pyrite, but it rarely forms crystals, has a specific gravity of 4.6, hardness of 3.5 to 4.5, and is slightly attracted by the magnet. It carries up to 30% sulphur, but so far has not been worked much for that element.

As a rapid and accurate method of estimating the sulphur available in a sample of pyrite, the following is the procedure: Place 5 grams of pyrite in a porcelain boat in a combustion tube, heat to redness, pass oxygen * over until combustion is complete, and absorb the gas formed in 30 cubic centimeters of a solution of bromine in a mixture of equal parts of hydrochloric acid (specific gravity, 1.1) and water, in potash (or preferably nitrogen) bulbs. Wash out the solution into a beaker, boil, precipitate by boiling solution of barium chloride, cool, filter, and wash, dry and ignite the barium sulphate.

Talc or Soapstone, called ' steatite ' when massive, is a hydrated silicate (56%) of magnesia (24%) from which the water is only driven off at a high temperature. It usually occurs in foliated laminar masses, like mica, but differs from the latter in not being elastic, in being softer and readily marked by the nail, in yielding an unctuous-feeling powder, and in not containing alumina as an essential ingredient. Deposits are to be found in serpentine and schists. The laminated variety of talc has been adopted by mineralogists as representing 1 in the scale of hardness; its specific gravity is 2.7. The color is white, sometimes tinged with green, and the luster pearly. When heated in a matrass, it undergoes no appreciable loss of water or transparency; when subjected to a high heat it exfoliates and hardens, but does not melt. Acids have no effect upon it, either after or before ignition. Talc is quarried and employed for various purposes, such as mixing with clay to increase the translucency of fin-

* The oxygen should be prepared from pure potassium chlorate in glass vessels, or at any rate in an iron one, kept especially for the purpose, and the gas should be passed through a strong solution of potash in the bulbs, through a U-tube containing calcium chloride, and lastly, either through another calcium chloride tube or, preferably, over phosphoric anhydride before use.

ished porcelain; when powdered it is used for diminish-
ing the friction of machinery, and as a basis for cosmetic
powders, which consumes large quantities. Pencils are
made from it for removing grease from silk and cloth, and
for marking the patterns of clothes; the latter is called
' French chalk.'

Tantalum. This is a rare metal, existing as a tantalate
of iron and manganese, containing 70% tantalum. It is
a heavy mineral, being 5.3 to 7.3 in the gravity scale, and
with a hardness of 6. Its color is iron-black; streak is
dark brown to black; and luster is sub-metallic. A little
has been found in California, South Dakota, and Virginia.
Western·Australia has been the main producer. There
it occurs as stibio-tantalite, a tantalate of antimony, in
pegmatite veins associated with tin. Some ore has been
found in Rhodesa also. Principal use of tantalum is in
electric lamps.

CHAPTER XVI

ABRASIVES, GEM-STONES, AND USEFUL ROCKS

Although there are many varieties of gems and precious stones in the United States, systematic mining for them is carried on only at a few places, and the annual output is very small ($319,000 in 1913 and $106,000 in 1918, Montana and Nevada accounting for 70%), but importations are valued at up to $40,000,000 a year. Not many persons are familiar with the appearance of gem-stones in their native state, so that quartz pebbles are often mistaken for rough diamonds, garnets for rubies, ilmenite for black diamonds, and so on. It is probable that many valuable stones have escaped notice.

Many gems are of comparatively little value, so that it is not always profitable to spend much time searching for them, unless a large quantity is found, as the costs of cutting and polishing is an important factor in their valuation. Many transparent and translucent kinds of quartz, colored by metallic oxides, are examples of such. It is so easy to prospect a stream in a country of crystalline, igneous, or metamorphic rocks, that a search for precious stones and gems of all kinds should be made much more frequently than is usually done. The precious varieties may be associated with all sorts of worthless specimens, all of which, though impure in quality, may really be sapphires, spinels, chrysoberyls, tourmalines, or zircons.

Though many are translucent rather than transparent, many dark in outward appearance, and all more or less water-worn with surfaces not at all glass-like, and the majority not apparently transparent or translucent unless held up to the light, yet a good specimen may be found among them. A knowledge of the general appearance of impure stones is of as much importance as that of the good ones, in aiding search for the latter.

Diamonds and gold are often found in the same alluvial deposit, and such ground should therefore be examined for this stone. The specific gravity of the diamond (3.5), — quartz or most pebbles (2.6).— and that of gold (16 to 19) are so different that it does not follow that these two minerals will always be found close together in a stream bed.

While certain characteristics of precious stones, such as hardness and specific gravity, may be useful to the prospector, it is not always an easy matter to distinguish a certain stone from one that may be similar in appearance though perhaps of much less value.

Diamond. Most of the diamonds of today come from the great deposits of blue and yellow ground at Kimberley and Pretoria, the alluvial areas in Orange River Colony, and those of southwest Africa. In September, 1919, it was reported that diamonds had been discovered in quartz gravels of a stream in West Africa.

In Borneo, the diamond is associated with magnetic iron ore, gold, and platinum, in deposits consisting of serpentine and quartz fragments, as well as marl.

The province of Minas Geraës, Brazil, is rich in diamonds, the most important occurrence being at Sao Joao do Barro, where they are found in an entirely weathered talcose-slate. In other parts of the same country the diamond is also obtained from a conglomerate of white quartz, pebbles, and light-colored sand, sometimes with yellow and blue quartz and iron sand. The province of Bahia produces the ' carbonado ' or black diamond. It is an allotropic form of carbon closely related to the diamond, and is found in small, irregular, crypto-crystalline masses of a dark gray or black color. Although its density is not so great as that of the diamond, it is much harder; in fact, it is the hardest substance known. At first it was used only in cutting diamonds, but with the ' bort,' another similar stone, is now extensively used in diamond-drills for exploring deep ground. The stones are set in a crown, which is worth up to $1200.

In South Africa, the diamond is associated chiefly with garnet and titanic iron ore, as well as with quartz opal, calc-spar, and more rarely with iron pyrite, bronzite, smaragdite, and vaalite. In the diamondiferous sands are topaz, garnet, bronzite, ilmenite, quartz, tremolite, asbestos, wallastonite, vaalite, zeolite, iron pyrite, brown iron ore, calc-spar, opal, hyalite, jasper, agate, and clay. The principal rocks are serpentine, eclogite, pegmatite, and talcose-slate. At De Beers' Kimberley mine, the diamond-bearing ground forms a ' pipe ' or ' chimney ' surrounded by formations totally different from the pay-rock. The encasing material is made up of red, sandy soil on the surface, below which is a layer of calcareous tufa, then yellow shale, then black shale, and below this, hard igneous rock. The formation consists of ' yellow ground ' (really the decomposed ' blue ground '), which is comparatively friable; and deeper down the ' blue ground ' (hydrous magnesian conglomerate) which needs blasting. The blue ground is of a dark bluish to a greenish-gray color, and has a more or less greasy feel. With it are mixed parts of boulders consisting of serpentine, quartzite, mica-schist, chlorite-schist, gneiss, granite, etc. All this blue ground has evidently been subjected to heat. Diamonds are formed by heat, and sudden cooling and pressure of rocks, this causing crystallization. The gems are in the matter that binds these rocks, not in the rocks themselves. The blue ground is spread out on the surface to weather, after which it is easily washed, and the diamonds saved on grease tables.

Diamonds are also found in the Urals of Russia, New South Wales, Australia (where they are recovered from gravel being washed for tin), and in the United States.

Diamonds have been found in Californian placer districts since the fifties, but their total is small. Their origin is believed to be the basic igneous rocks from which the serpentines of the gold districts have been derived. Geologists consider that since so many associations of the diamond are present in North Carolina, they have hopes of their being found there eventually. In Wisconsin, diamonds have been found in glacial deposits with quartz, garnet, ilmenite, and magnetite.

The natural surface of the diamond is often unequal; its sides are lined, somewhat convex, and generally appear dulled, or as they are commonly called, 'rough.' The diamond breaks regularly into four principal cleavages. It does not sparkle in the rough, and the best test is its hardness and its becoming electrified when rubbed before polishing. The color of the diamond varies through all tones of the color-scale — from colorless through all shades of yellow, red, green, and blue, to intense black.

Some colorless diamonds acquire a reddish shade on heating, which disappears on cooling. Blue stones are highly prized, and often bring better prices than the colorless.

The specific gravity of the pure diamond varies from 3.5 to 3.6; that of the black diamond is from 3.012 to 3.255.

One of the most beautiful qualities of the diamond is its power of refraction; that of·water is 0.785; ruby, 0.739; rock crystal, 0.654; and the diamond, 1.396. The refraction of the diamond is single in the entire crystals; when broken it possesses double, but imperfect, refraction in thin layers.

The value of the diamond is dependent on its color, size, and the finish or cut given during polishing. Perfectly colorless stones bring the highest price, followed by those with a reddish, greenish, and bluish shade, which, however, are rare. Yellowish diamonds are of less value, the price paid for them being lower the more the yellow color shades off into brown.

The largest diamond ever found was the Cullinan, discovered at shallow depth in the Premier mine at Pretoria, Transvaal, South Africa, in 1905. It weighed 3024 carats (1 carat = 3.2 grams) or 1⅜ pounds. It was considered to have been part of an original stone, and late in 1919 it was announced that the other part had been found. One weighing 442 carats was found at Kimberley in 1917. Of course, such diamonds are extraordinary and of rare occurrence.

Corundum. This mineral includes some of the most important precious stones, such as the blue crystalline variety or sapphire, the red or ruby; the light-yellow or Oriental topaz; the bright-green Oriental emerald; and the bright-violet Oriental amethyst. One variety exhibits a six-rayed star inside the prism and is called the asterias.

Corundum mining has increased within the last few years. In Ontario, Canada, it is found in syenite; and in North Carolina and Georgia, in a gneiss or a quartz-schist. American deposits are practically limited to these two places. The output of corundum as a gem in this country was $40,000 in 1918, but as an abrasive 17,135 tons in 1917.

The emery bed at Chester, Massachusetts, has furnished a large quantity of this mineral.

Corundrum was formerly regarded as occurring sparingly in nature, and in only a few types of rocks, but it is now known to occur rather widely, and instead of being in quantity in the basic magnesian or periodite rocks only, it has been found in abundance in syenite, gneiss, and schist. Although found in many of the crystalline rocks it has been observed as a rock constituent in only a few of them. In some cases it is an original constituent of the rock, in others it has been formed later, during the process of metamorphism.

Corundum, as mined for abrasive purposes, occurs as sand, crystal, or gravel, and block corundum. Sometimes all three types are found in the same deposit. The first consists of small grains or fragments of the mineral scattered through a bed. The second consists of crystals up to three inches in length. The block variety often occurs in masses of almost pure corundum from 10 to 1000 pounds in weight. Again, it is found in large masses intimately associated with hornblende, or feldspar, making a rock which is tough and difficult to work.

It is the hardness of corundum that makes it of so great a value as an abrasive. Next to the diamond it is the hardest mineral known, being 9 on the scale. Its specific gravity is 3.9 to 4.2; luster, glassy, sometimes pearly; and fracture, uneven or conchoidal. It is infusible before the blowpipe, and not affected by acids nor heat. It is an oxide of aluminum, containing a small percentage of other constituents, principally silica, water, and ferric oxide.

Emery is a granular, impure form of corundum, consisting of a mechanical mixture of corundum and magnetite or hematite. It is of great commercial value as an abrasive, though carborundum (an artificial abrasive made from silica and carbon in the electric furnace) is displacing it largely. The chief foreign sources of emery are the Greek island of Naxos, and Asiatic Turkey, India, Canada, and South Africa.

Sapphire. This is the blue variety of corundum in its purest crystalline state. Its general composition is alumina, 92%; silica, 5.25%; oxide of iron, 1%. The color most highly valued is a highly transparent bright blue. More frequently the color is pale blue, passing by paler shades into perfectly colorless varieties. The paler gems are frequently marked by dark blue spots and streaks, which detract from their value; but these lose their color when subjected to great heat, a fact which has some times been taken advantage of by unscrupulous dealers to pass them off as diamonds.

The principal form of the sapphire is an acute rhomboid, but it has many modifications. On being broken it shows a conchoidal fracture, seldom a lamellar appearance.

In the United States there are two areas of importance as producers of sapphires. One is in Macon county, North Carolina, where the mineral occurs with spinel, tremolite, tourmaline, magnetite, rutile, chromite, olivine, and mica, in gneiss. The other district is in Montana; which State yielded $47,000 of sapphires in 1918. Near Helena is a glacial moraine known as El Dorado bar, and in this sapphire has been found with topaz, garnet, cassiterite, quartz, and cyanite. In addition it has been found in place in Montana in a dike with pyrope, at Yago creek, near Judith river, of a fine corn-flower blue. At Santa Fe, New Mexico, in southern Colorado, and in Arizona, sapphires exist in the sand associated with peridot, pyrope, and almandine garnet.

Probably the best sapphires in the world come from Queensland, Australia.

Ruby. The ruby is the red variety of corundum and in composition varies from almost pure alumina to a compound containing 10 to 20% of magnesia, and always about 1% of oxide of iron. The ruby is subdivided into several varieties according to color, which in its turn is affected by mineral composition, spinel ruby occurring in bright red or scarlet crystals; rubicelle, orange-red; bala ruby, rose red; almandine ruby, violet; chlorospinel, green; and pleonast, is the name given to dark varieties.

The crystals are usually small, and when not defaced by friction have a brilliant luster, as has the lamellar structure, with natural joints which it shows on being broken. It exhibits various degrees of transparency. The color most valued is the intense blood-red or carmine of the spinel ruby. When the color is a lilac-blue, the specimen was formerly known as the Oriental amethyst, and was regarded as a connecting link between the ruby and the sapphire. In the United States, ruby is found in various localities. It occurs in gneissic and metamorphic rocks and in granular limestone. In North Carolina, ruby of gem quality has been found at Corvee creek, a tributary of the Little Tennesse river, in a decomposed garnetiferous basic rock. Many much-weathered specimens have been found, which have led to the conclusion that rubies of large size have been formed there. The associated minerals are garnet, spinel, monazite, rutile, ilmenite, different micas, staurolite, and gold. The stones often show inclusions, sometimes so minute as to give the gem a 'sheen.'

The ruby ranks above the diamond in point of value for good stones of rich deep color.

The garnet is sometimes mistaken for the East Indian ruby, which is the most precious variety, but the garnet is isometric, and even when cut and mounted may be distinguished from the oriental ruby by the superior hardness of the latter, which is 9 on the scale, while garnet is 6.5 to 7.5. A garnet of the best kind, if worn will, under a strong lens, show the lines of wear, especially on the edges, which are absent in the true oriental ruby. Burma is the principal source of rubies.

Topaz is composed of silica, alumina, and fluorine. The fluorine may be detected before the blowpipe in an open tube by powdering a little of the topaz and mixing it with a small quantity of microcosmic salt (a salt of phosphorus). The heat of the blowpipe will liberate the fluorine, and its strong pungent odor, as well as its corrosion of the tube will prove its presence. With a cobalt nitrate solution on charcoal it gives a fine blue color, which shows alumina. This is the best test of the topaz, as the color of the mineral is not always the same, nor is it always perfectly transparent.

Topaz crystallizes in the orthorhombic section of the hexagonal or fourth system. The finest are generally in prismatic form, showing a flat plane at the extreme end, even when the end of the crystal has several inclined faces. The crystals break easily across with smooth, brilliant cleavage. Topaz is transparent or semi-transparent; white, yellow, greenish, bluish, or pink; has a glassy luster; specific gravity of 3.4 to 3.65 and hardness of 8. It scratches quartz, and is scratched by sapphire. It is infusible, but often blistered and altered by heat. When smooth surfaces are rubbed on cloth they become strongly electrified, and can attract small pieces of paper; but rough surfaces do not show this. The brilliant cleavage of topaz distinguishes it from tourmaline and other minerals. Topaz occurs in gneiss or granite with tourmaline, mica, beryl; also cassiterite or tin-stone, apatite, and fluorite. The white topaz resembles the diamond, but unlike the latter it can be scratched by sapphire. The pale blue variety is of value for cutting into large stones for brooches; specimens are occasionally found weighing several pounds. Topaz of a beautiful sherry color is found in Brazil. Such stone when heated becomes pink, and is known as burnt topaz. The yellow varieties are cut as gems. Although not very valuable, they are brilliant and look well.

Topaz is usually found, when in place, in attached crystals in cavities in granites. It is also frequently associated with beryl, tourmaline, and feldspar. Topaz resists most weathering, hence is often found in rolled pebbles in the detritus of granite and other rocks.

In the United States, yellow crystals are found in Connecticut, blue ones in granite in Maine, colorless ones in Utah, and both colorless and pale blue occur with Amazon stone at Florissant in the Pike's Peak district of Colorado.

Beryl or Emerald is composed of silica, alumina, and beryllium or glucinum. It is almost always found in distinct crystals, and usually in forms easy to recognize. The crystals are hexagonal prisms, usually green, transparent, or opaque. Luster — glassy; fracture — uneven; specific gravity — 2.63 to 2.80; and hardness — 7.5 to 8. It is infusible or nearly so, but becomes cloudy by heating. Beryl is found in granite rocks with feldspar and and quartz, and is valuable for jewelry when transparent and rich grass-green (emerald) or sea-green (aquamarine).

Both emerald and aquamarine are found in Alexander county, North Carolina. Green and golden beryl occur in Oxford county, Maine, and beryl of a sapphire blue is found at Royalston in Massachusetts. Emerald also occurs at Haddam in Connecticut. On Mount Antero, in Colorado, fine aquamarines are found associated with phenacite.

A productive emerald mine was that of Muso, in New Granada, Mexico. The emerald occurs in veins and cavities in a black limestone containing fossil ammonites. The limestone also contains within itself minute emeralds and an appreciable quantity of glucina. When first obtained the emeralds from this mine were soft and fragile; the largest and finest emeralds could be reduced to powder by squeezing and rubbing them with the hand. After exposure to the air for a little time they become hard and suitable for the jeweler.

Phenacite is a silicate of beryllium or glucinum. Its hardness is about the same as topaz, and its specific gravity 3.4 to 3.6. It occurs in glassy rhombohedral crystals, and its hardness, beautiful transparency, and color make it valuable for cutting as a gem, since it is capable of extreme polish. Phenacite has been found at Pike's Peak, Colorado, in crystals of sufficient size and quality to yield fair gems. It also occurs at Topaz Butte with topaz and Amazon stone in granite; also at Mt. Antero, in Colorado with quartz and beryl.

Zircon is composed of silica and zirconia. It is found in square prisms terminated by pyramids, and in octahedrons, but often also in pebbles and grains. It is transparent or opaque; wine or brownish-red, gray, yellow, white color; and glassy luster. The fracture is usually irregular, but in one direction it can be split so as to exhibit a smooth even cleavage-face having an adamantine luster like the diamond. Its specific gravity is 4.68 to 4.70, and hardness, 7.5. It scratches quartz, and is scratched by topaz. Infusible; the red varieties, when heated before the blowpipe, emit a phosphorescent light, and become permanently colorless. Zircon is found in syenite, granite, and basalt. In some regions it occurs so abundantly that when the rock has been eroded down, it is left unaltered in considerable quantities. It may then be obtained by washing the gravel in the same way that placer gold is saved. Clear crystals are used in jewelry, as watch jewels, and as an imitation of the diamond. It may be distinguished from the latter by its inferior hardness, and in not becoming electrified by friction so readily. Fine crystals are obtained in New York and Canada; and good specimens also come from North Carolina and Colorado. The most recent uses of washed, iron-free zircon is for abrasives, alloys, enamels, glasses, pigments, refactories, and salts.

Garnet is composed of silica, alumina, lime, iron, magnesium, manganese and chromium. It is found nearly always in distinct crystals, and as these are commonly isolated and scattered through the rock, it is not difficult to recognize them. Where garnet is a contact mineral formed by the intrusion of igneous rock into limestone and other rock it is often found in large crystals. The crystals are usually twelve-sided, having the form of a rhombic dodecahedron. They are transparent or opaque; generally red; also brown, green, yellow, black, white. Luster — glassy or resinous; fracture — conchoidal or uneven; specific gravity — 3.5 to 4.3; and hardness — 6.5 to 7.5, cannot be scratched with a knife. Fusible with more or less difficulty. Red varieties impart a green color to borax bead owing to presence of chromium. Garnet usually occurs in crystals scattered through granite, quartzite, gneiss or mica-schist, also in crystalline limestone; with serpentine or chromite; also in some volcanic rocks. It is generally a product of metamorphism. Garnet also occurs in beach sand. Fine-colored transparent varieties, such as cinnamon stone and almandite are used in jewelry. Garnets found in Arizona, New Mexico, and southern Colorado, and there called 'rubies,' are as fine as those from any other locality, the blood-red being the most desirable. Fine crystals of cinnamon stone, cinnamon garnet, or essonite have been found in

New Hampshire, Maine, and at many other points in the United States. Alaska and California produce good garnets. Besides its value as a gem, garnet finds its greatest use as an abrasive. During 1918, New York, New Hampshire, and North Carolina contributed most of the 4127 tons mined in the United States.

Tourmaline is composed of silica, alumina, magnesia, boracic acid, fluorine, oxides of iron, lime and alkalies. It is found in prisms with three, six, nine, or more sides, furrowed lengthwise, terminating in low pyramids. Commonly black and opaque, rarely transparent, and of a rich red, yellow, or green color. Its luster is glassy; fracture, uneven; specific gravity, 2.98 to 3.20; and hardness, 7 to 7.5. It cannot be scratched with a knife. When the smooth side of a prism is rubbed on cloth it becomes electrified, and can attract a small piece of paper. Tourmaline occurs in granite and slate. Only the fine-colored transparent varieties, which are used as gems and for optical purposes, are of value. Probably the finest tourmaline in the world are the red and green varieties in San Diego county, California. There it is found in a series of pegmatite veins consisting mainly of feldspar, quartz, and mica, cutting through diorite.

Epidote is a silicate of alumina, iron, and lime, but varies rather widely in composition, especially regarding the relative quantities of alumina and iron. It is usually found in prismatic crystals, often very slender and terminated at one end only; they belong to the monoclinic system. Luster is vitreous; and color, commonly green, although there are black and pink varieties. Epidote is found in many parts in the United States, in large crystals ranging from brown to green in color, but as a rule they are only translucent or semi-opaque, though some stones of considerable value and great beauty have been found in Rabun county, Georgia.

Opal is composed of silica and water. It is never found in crystals; but only in massive and amorphous form. The fracture is conchoidal; specific gravity, 2.2; and hardness, 5.5 to 6.5. It can be scratched by quartz and thus be distinguished from it. It is infusible and generally milk-white. The most important variety is that called precious opal,' which exhibits a beautiful play of colors and is a valuable gem. One kind of precious opal with a bright red flash of light is called 'fire opal,' and another kind is the ' harlequin opal.' Common opal does not exhibit this play of colors, and it varies widely in color and appearance. ' Milk opal,' as one variety is called, has a pure white color and milky opalescence, while ' resin opal ' or ' wax opal ' has a waxy luster and yellow color. ' Jasper opal,' is intermediate between jasper and opal; ' wood opal ' is petrified wood, in which the mineral material is opal instead of quartz. Opal is commonly met with in seams of certain volcanic rocks; sometimes it occurs in limestone and also in metalliferous veins. Precious opal is rare in the United States, though some of high value is said to have been found near John Davy's river, Oregon.

The precious opal found in Hungary occurs in fissures in a weathered and andesitic lava with other forms of opal. Hungarian opals show the finest fire, and their colors deteriorate least with exposure.

In Mexico, precious opal has been mined in the State of Queretaro, in volcanic rock and associated with other forms of opal. The colors are intense, but in larger

patches than the Hungarian specimens show, and the colors do not change so much when the stone is moved.
Some opal has been mined in Humboldt county, Nevada, near the Californian and Oregon lines.
Probably the finest opal in the world comes from White Cliffs, near Broken Hill, New South Wales, Australia. The gem from there has the most wonderful coloration, including the fire opal. A valueless stone called ' false opal ' is found there, but the miners soon learn how to tell it from real opal.

Turquoise is a hydrous phosphate of aluminum, containing also a little copper phosphate,— probably the source of the color — which in the most precious variety is robin's-egg blue, and bluish-green in less highly-prized varieties. It occurs only in compact massive forms, filling seams and cavities in volcanic rock. Specific gravity is 2.6 to 2.8. Turquoise has been found in the Holy Cross mining region, 30 miles from Leadville, Colorado, and a number of mines have been opened at Los Carillos and in Grant county, New Mexico. The latter mines produce stones having a faint greenish tinge, due either to a part change or metamorphism, which has taken place while the turquoise was in the rock, or it may be a local peculiarity. Turquoise occurs also in Arizona and in Southern Nevada. At the latter place it is found in veins of small grains in a hard shaly sandstone. The color of this turquoise is a rich blue, almost equal to the finest Persian, and the grains are so small that the sandstone is cut with the turquoise in it, making a rich mottled stone for jewelry.

Agate is found in almost every part of the world. Its specific gravity varies from 2.58 to 2.69. The agate, properly so called, is naturally translucent, less transparent than crystalline quartz, but yet less opaque than jasper. It is too hard to be even scratched by rock crystal. It takes a good polish. It is never found in regular forms, but always either in nodules, in stalactites, or in irregular masses. Eye agates consist of those parts of the stone in which the cutting discovers circular bands of small diameter arranged with regularity round one circular spot. These circles are frequently very perfect. The first round is white; the second, black, green, red, blue, or yellow; and the rarest are those whose circles are at equal distance from the center. Moss agate contains brown-black, moss-like, or dendritic forms distributed rather thickly through the mass. These forms consist of some metallic oxide — such as manganese. Of all the American stones used in jewelry there is no other of which so much is sold as the moss agate. The principal sources of supply are Utah, Colorado, Montana, and Wyoming.

Chalcedony is a semi-transparent variety of quartz of a waxy luster, varying in color from white through gray, green and yellow to brown. It is translucent to opaque. It occurs in stalactite, reniform or botryoidal masses, which have been formed in cavities in greenstones and others of the older rocks. Into these cavities, as into miniature caverns, water-holding silicious matter has penetrated and deposited its solid contents, consisting almost exclusively of silica tinged by the presence of other minerals. Some of these cavities are several feet in diameter, and besides the coloring of the encircling mass there are often, in the interior of the concretions in them, cavities or central nuclei that contain sometimes as many as 24 different substances, such as silver, iron pyrite, rutile, magnetite, tremolite, mica, tourmaline, topaz, with water, naphtha, and air.

A comparatively recent use for chalcedony is in the tube or grinding mills at gold and silver mines and at cement plants. The rock is broken into small pieces and rounded into pebbles up to 5 inch size for this purpose. Such pebbles have replaced foreign importations. A deposit near Manhattan, Nevada, and one at Sioux City, Iowa, is producing this product.

Carnelian is chalcedony colored by the oxide of iron — hematite. It is sometimes called ' sard.' It has the same properties as chalcedony, and occurs either as an ordinary agate, or in fissures as vein agate. Although of wide distribution, only two localities, both in India, are known. The name carnelian was given to the stone on account of its flesh color.

Chrysoprase is a variety of chalcedony colored green by oxide of nickel. In a warm dry place its color is destroyed, but it can be restored by keeping it damp.

Jasper is quartz rendered opaque by clay, iron, and other impurities. It is of a red, yellow or green color. Sometimes the colors are arranged in ribands, or in other fantastic forms. It is used for ornamental work.

Wood Jasper is a fossil wood silicified. It is found at Chalcedony Park, Arizona, and at Yellowstone Park, Wyoming.

Bloodstone or heliotrope is green jasper, with splashes of red resembling blood spots.

Rock crystal is pure, transparent, colorless quartz, and is found at a great many places in the United States. In Herkimer county, at Lake George, and throughout the adjacent regions in New York state, the calciferous sandstone contains single crystals, and at times cavities are found filled with doubly terminated crystals, often of remarkable perfection and brilliancy. These are collected, cut and, often uncut, are mounted in jewelry and sold under the name of ' Lake George diamonds.'

What is known as ' brilliant quartz ' is used for glass and fused quartz ware, used largely in laboratories now.

Amethyst is a transparent variety of quartz of a rich violet or purple color due to oxide of manganese. It crystallizes in the form of a hexagon, terminated at the two heads by a species of cone with six facets. These crystals are often in masses, and the base is always less colored than the top. Amethysts are generally found in metalliferous mountains, and are always in combination with quartz and agate. They are found in many parts of the United States, for instance, near Greensboro, in North Carolina, and in the Lake Superior region, especially in the northwest, but not in so fine or large specimens as in Ceylon or Siberia.

Rose quartz is pink, red, and inclining to violet-blue in color. It usually shows a vitreous luster, and small conchoidal fracture. As a rule it is not crystallized, and but rarely transparent. It is liable to fade on exposure, though it may to some degree be restored by moistening the specimen.

Yellow Quartz or citrine or false topaz occurs in light-yellow translucent crystals. It much resembles yellow topaz in color, and hence is often called ' Occidental topaz ' or ' Spanish topaz.' It is often set and sold for topaz, but may be distinguished from it by its want of cleavage and by being softer.

Smoky quartz or cairngorm varies in color from a

pale sherry tint through all degrees of smoky brown to almost black. It occurs in crystals identical in all respects, except color, to rock crystal. Its commonest occurrence is in fissures in granite and allied rocks, sometimes in spaces in the outer parts of a granite mass, probably due to shrinkage on consolidation; in such cavities sometimes associated with beryl, topaz, and crystals of feldspar.

Onyx and Sardonyx. A variety of quartz having a regular alternation of strata more or less even, and variously colored in black, white, brown, gray, yellow, and red. When the onyx has one or two strata of red carnelian, it is more valued and takes the name of sardonyx. In the onyx the dark strata are always opaque and contrast strongly with the clear, which, when thinned, become almost translucent. One of the important limestones is onyx marble, and the two onyx should not be confused.

Cat's Eye consists of a quartz mixed with parallel fibers of asbestos and amianthus. It is found in pebbles and in pieces more or less rounded; it has a concave fracture; is translucent and also transparent at the edges. It has a vitreous and resinous light. It is generally either green, red, yellow, or gray. It marks glass. Its specific gravity is from 2.56 to 2.73. When exposed to a great heat it loses luster and transparency, but does not melt under the blowpipe unless reduced to minute fragments.

Many other gem-stones are known in the United States, but those given are of most importance. A text-book on gems by G. F. Kunz, of Tiffany & Co., New York, is recommended for further study of this subject.

APPENDIX

WEIGHTS AND MEASURES

One of the subjects that we learn at school and are liable to forget, is weights and measures; so it is sometimes handy to have them for reference, and on the following pages are given those commonly used:

American and British weights and measures are based upon the weight of a cubic inch of distilled water at 62° Fahrenheit, and 30 inches height of the barometer, the maximum density. This was decided early in the 19th century to be 252.458 grains but it has been proved since that a cubic inch of water at the temperature of maximum density is 252.286 standard grains.

No. 1.— Length

12 inches	=	1 foot.
3 feet	=	1 yard.
5½ yards	=	1 rod, pole, or perch (16½ feet).
4 poles or 100 links	=	1 chain (22 yards or 66 feet).
10 chains	=	1 furlong (220 yards or 660 feet).
8 furlongs	=	1 mile (1760 yards or 5280 feet.)
80 chains	=	1 mile.
100,000 sq. links or 10 sq. chains	=	1 acre.

A span = 9 inches; a fathom = 6 feet; a league = 3 miles; a geographical mile = 6082.66 feet, same as nautical knot, 60 being a degree, that is 69.121 miles.

PARTICULAR MEASURES OF LENGTH

A point, ½ of an inch.
A line, ¹⁄₁₂ of an inch.
A palm, 3 inches.
A hand, 4 inches.
A link, 7.92 inches.

A pace, military, 2 feet 6 inches.
A pace, geometrical, 5 feet.
A cable's length, 120 fathoms.
A degree (average), 69⅛ miles.

No. 2.— SURFACE MEASURE

144 square inches	=	1 square foot.
9 square feet	=	1 square yard.
30¼ square yards	=	1 pole, rod, or perch (square).
16 poles (square)	=	1 chain (sq.) or 484 sq. yd.
40 poles	=	1 rood (sq.) or 1210 sq. yd.
10 chains or 4 roods	=	1 acre (4840 sq. yd.)
640 acres	=	1 sq. mile.

No. 3.— SURFACE MEASURE IN FEET

9 square feet	=	1 square yard.
272¼ " "	=	1 pole, rod, or perch.
4,356 " "	=	1 square chain.
10,890 " "	=	1 square rood.
43,560 " "	=	1 acre.
27,878,400 " "	=	1 square mile.

No. 4.— SOLID MEASURE

·1728 cubic inches	=	1 cubic foot.
27 cubic feet	=	1 cubic yard.

16½ feet long, 1 foot high, and 1½ feet thick = 1 perch stone = 24¾ cubic feet.

No. 5.— TROY WEIGHT

Platinum, gold, silver, and some precious stones are weighed by Troy weight; diamonds by carats of 3.12 grains each.

24 grains	= 1 pennyweight.
20 pennyweights	= 1 ounce (480 grains).
12 ounces	= 1 pound (5760 grains).

No. 6.— Avoirdupois Weight

27.34 grains	= 1 dram.
16 drams	= 1 ounce (437½ grains).
16 ounces	= 1 pound (7000 grains).
112 pounds	= 1 hundred-weight.
100 pounds	= 1 hundred-weight.
20 hundred-weight	= 1 ton (long ton) (2240 pounds).
20 hundred-weight	= 1 ton (short ton) (2000 pounds.)

No. 7.— Weight by Specific Gravity

Frequently the weight of a mass is required where it is inconvenient, or, perhaps, impossible to use scales. The following method may be sufficiently accurate:

Find the average specific gravity of the mass either by actual weight of a piece, or by the following calculation; then measure the cubic contents of the mass as nearly as possible and multiply by the weight of a cubic foot. Thus, a block of limestone, such as good marble, measures 40 cubic feet. Its specific gravity is 2.6, that is, it is 2.6 times as heavy as a cubic foot of water, which weighs 62.5 pounds. Therefore

$$\begin{array}{r} 62.5 \\ 2.6 \\ \hline 3750 \\ 1250 \\ \hline 162.50 \end{array}$$

A cubic foot of marble weighs 162.5 pounds, and the 40 cubic feet will weigh

$$\begin{array}{r} 162.5 \\ 40 \\ \hline 6500.0 \end{array}$$

or, about 3¼ tons. Of course, all rock masses have not plane sides, so this irregularity requires some calculation and allowances that must be made, which can easily be done with a little consideration.

When greater accuracy of specific gravity and of bulk is desired for small masses, and no scales are at hand, the following plan may be adopted satisfactorily. Fill a tub, barrel or large box with rain water, after having inserted a tube or piece of tin pipe into the upper edge. Pour in more water until it will hold no more without running out of the spout. Put in the rock and catch all the water that flows out of the pipe. Next measure the overflow; this represents the exact cubic measure of the rock.

```
1 gallon contains  .........231 cubic inches.
1 quart     "      .........87.75 or 57¾ cubic inches.
1 pint      "      .........28.87 or 28⅝   "     "
1 gill      "      ......... 7.21 or  7⅕   "     "
      See No. 8
```

Suppose the overflow was 8 gallons, 1 quart, 1½ gills, and that the specific gravity of the rock or ore was 6.5 by the calculation below. Then the mass will cause an overflow of 1936.99 cubic inches, and this is 208.99 more than one cubic foot, or about 1.120 of a cubic foot for the mass.

Since 6.5 was the specific gravity of the ore, 6.5 × 62.5

pounds = 406.25, which would be the weight of a cubic foot of the ore; and 406.25 × 1.120 = 455 pounds, the exact weight of the mass put into the water.

No. 8.— Miscellaneous Weights and Measures

One cubic foot of water is equal to 7.475 United States gallons of 231 cubic inches each, or 7½ gallons nearly; or 6.2321 Imperial gallons of 277¼ cubic inches each. This is important in the construction of tanks, reservoirs, and other containers, where contents, weight, and pressure are to be considered.

It should be remembered that, although the British imperial gallon is 277¼ cubic inches or 10 lb. avoirdupois of distilled water at 62° F., barometer 30 inches, and equal to 277.274 cubic inches, the United States standard gallon is 231 inches, or 58328.886 grains, or 8.3326 lb. of distilled water at maximum density. This is almost exactly equal to a cylinder 7 inches diameter, by 6 inches high. The beer gallon (a thing of the past in America) equals 282 inches.

One gallon = 8.3326 lb.; one quart = 2.0831 lb.; one pint = 1.0415 lb.; one gill = 0.2604 lb., U. S. Standard measure. One cubic foot of water = 62.278 lb., British weight.

The Metric System

This system of weights and measures was devised by the French late in the 18th century, and has since been adopted by all of Europe, Mexico, Central and South America, and several other nations. Although the metric or decimal system is used by the British Empire and the United States to a limited extent, it does not gain much headway for general purposes, in spite of strong support for it by certain authorities. Calculations in assaying, chemistry, and other sciences are founded on metric measures, hence it is essential that all should understand them.

The metric system is based upon the (assumed) length of the fourth part of a terrestrial meridian, that is, the distance from the equator to one of the poles. The tenmillionth part of this arc was chosen as the unit of measures of length, and called a *mètre*. The cube of the tenth part of the mètre was adopted as the unit of capacity, and denominated a *litre*. The weight of a litre of distilled water at its greatest density was called a *kilogramme*, of which the thousandth part, or *gramme*, was adopted as the unit of weight. The multiples of these, proceeding in decimal progression, are distinguished by the employment of the prefixes *deca, hecto, kilo* and *myria*, from the Greek, and the sub-divisions by *deci, centi*, and *milli*, from the Latin. To be clearer, all of the weights and measures are obtained by either sub-dividing or multiplying by 10. The termination ' re ' is now generally written ' er,' and the ' me ' is cut off gramme.

No. 9.—LENGTH.

Millimeter (¹⁄₁₀₀₀ of a meter)	=	0.03937 inch.
Centimeter (¹⁄₁₀₀ of a meter)	=	0.3937 inch.
Decimeter (¹⁄₁₀ of a meter)	=	3.937 inch.
Meter (the unit of length)	=	39.3708 inch or 3.2809 ft.
Decameter (10 meters)	=	32.809 ft. or 10.9363 yds.
Hectometer (100 meters)	=	109.3633 yards.
Kilometer (1000 meters)	=	1093.63 yd. or 0.62138 mile.
Myriameter (10,000 meters)	=	6.2138 miles.

SURFACE

Centiare (⅒ of an are or sq. meter) = 1.1960 sq. yd.

Are (unit of surface) = $\begin{cases} 119.6033 \text{ sq. yd. or} \\ 0.0247 \text{ acre.} \end{cases}$

Decare (10 ares) = $\begin{cases} 1196.033 \text{ sq. yards or} \\ 0.2474 \text{ acre.} \end{cases}$

Hectare (100 ares) = $\begin{cases} 11960.33 \text{ sq. yd. or} \\ 2.4736 \text{ acres.} \end{cases}$

SOLID MEASURE.

Decistere (⅒ of a stere) = 3.5317 cubic feet.
Stere (cubic meter) = 35.3166 " "
Decastere (10 steres) = 353.1658 " "

WEIGHT.

Milligram (⅟₁₀₀₀ of a gram) = 0.0154 grain.
Centigram (⅟₁₀₀ " ") = 0.1544 grain.
Decigram (⅒ " ") = 1.544 grains
Gram (unit of weight) = 15.432 grains.
Decagram (10 grams) = 154.432 grains.

Hectogram (100 ") = 1,544 grains $\begin{cases} 3.2167 \text{ oz. Troy} \\ \text{or } 3.5291 \text{ oz.} \\ \text{Avoir.} \end{cases}$

Kilogram (1000 ") = 32¼ oz. or 2.204 pounds.
Myriagram (10,000 grams) = 22.04 pounds.
1000 kilograms = 2204 pounds or 1 tonne.

No. 10.— SPECIFIC GRAVITY OF METALS, ORES, ROCKS, ETC.

Platinum 14–19
Gold 15.6–19.3
Mercury 13.59
Lead........................... 11.35–11.5
Silver 10.1–11.1
Copper 8.5–8.9
Iron when pure 7.3–7.8
Iron, cast, average .... 6.7; foundry 6.9 to 7

ORES: associated with gold and silver.

(Gold) Iron pyrite	4.8–5.2
Copper pyrite	4.0–4.3
(Silver) Galena	7.2–7.7
Glance (silver)	7.2–7.4
Ruby silver (dark)	5.7–5.9
" " (light)	5.5–5.6
Brittle silver (sulphide)	5.2–6.3
Horn silver	5.5–5.6

OTHER ORES.

Zinc blende	3.7–4.2
Mercury (cinnabar)	8.8–9.9
Tin, tinstone, cassiterite	6.4–7.6
Tin pyrite	4.3–4.5
Copper — Red or ruby copper	5.7–6.15
Gray	5.6–5.8
Black oxide	5.2–6.3
Pyrite	4.1–4.3
Carbonate (malachite)	3.5–4.1
Lead — Sulphide (galena)	7.2–7.7
Carbonate (white lead)	6.4–6.6
Zinc — Blende	3.7–4.2
Calamine	4.0–4.5
Iron — Hematite (red)	4.5–5.3
Magnetic	4.9–5.9
Brown hematite	3.6–4.0
Spathic (carbonate)	3.7–3.9
Pyrite (mundic)	4.8–5.2
Antimony — gray sulphide	4.5–4.7
Nickel — Kupfer nickel	7.3–7.5
Cobalt — tin-white	6.5–7.2
Glance	6.0
Pyrite	4.8–5.0
Bloom	2.9–2.9
Earthy	3.1–3.3

Manganese — black oxide 4.7–5.0
Wad, bog manganese 2.0–4.6
Bismuth — sulphide 6.4–6.6
Oxide 4.3

MINERALS OF COMMON OCCURRENCE.

Quartz 2.5–2.8
Fluorspar 3.0–3.3
Calc-spar 2.5–2.8
Barytes (barite) 4.3–4.8
Gneiss }
Granite } 2.4–2.7
Mica slate 2.6–2.9
Syenite 2.7–3.0
Cyanite 3.56–3.67
Greenstone trap 2.7–3.0
Basalt 2.6–3.1
Porphyry 2.3–2.7
Talcose slate 2.6–2.8
Clay slate 2.5–2.8
Chloritic slate 2.7–2.8
Serpentine 2.5–2.7
Limestone and Dolomite 2.5–2.9
Sandstones 1.9–2.7
Shale 2.8

Other minerals are mentioned in the text with their specific gravities.

No. 11.—CUBIC FEET IN A TON OF CERTAIN MATERIALS.

Earth	21 cubic feet.	Pit sand	22 cubic feet.
Clay	18 " "	River sand	19 " "
Chalk	14 " "	Marl	18 " "
Coarse gravel	19 " "	Shingle	23 " "
Quartz in place	13 " "	Chromite	10 " "

STANDARD VALUE OF GOLD IN DIFFERENT COUNTRIES

COUNTRIES	1000 (24 carats).	916.66 (22 carats).	900 (21.6 carats). .
England .... (one troy ounce)	£4 4 11½	£3 17 10	£3 16 6
United States...... (one troy ounce)	$20.67	$18.95	$18.60
France (Kilogram)..	Fr. 3,444.44	Fr. 3,157.40	Fr. 3,100
Germany " ..	Mk. 2,790	Mk. 2,474.16	Mk. 2,511

Pure gold is 24 carats or 1000 fine; standard gold
(British basis), 22 carats, or $\dfrac{22 \times 1000}{24} = 916.66$ fine.

MEASURE OF WATER

A miner's inch is a variable quantity of water, but
generally is the amount that will flow through an opening
one inch square under a six-inch head. This is 2274
cubic feet in 24 hours, or 655 gallons hourly, equal to
say 11 gallons a minute.

To obtain the pressure of water in a pipe, multiply
the head or fall in feet by 0.43. Thus: 100 feet head
$\times 0.43 = 43$ pounds per square inch.

UNIT SYSTEM OF BUYING AND SELLING ORES

All minerals and metals excepting gold are subject to
daily market fluctuations, although some of them, through
lack of demand or by being closely controlled, show little
variation over long periods. When ores are sold, the
value is determined from the metal contents as shown
by assay, and the current price for those metals. (The
complications of treatment charges, bonuses, and penalties
will not be considered here.) After these factors are
known, there must be some basis for calculations, so the

unit system is used, mostly for the non-metallic ores. A unit is a $\frac{1}{100}$ or 1% of a short (2000 lb.) or long (2240 lb.) ton, therefore it is equal to 20 or 22.4 lb. The short ton is used for most ores, iron and manganese being two of the few exceptions. The rule for finding the value of an ore by the unit method is to multiply the number of units or percentage of metal in it by the market quotation, which itself is based on a fixed percentage of metal, plus the demand and supply. For instance: we have some antimony ore assaying 30% or units of metal, and the price is 70 cents per unit, based on ore carrying 50% metal, the value is

$$30 \times 70 = \$21 \text{ per ton}$$

Conversely, if 30% antimony ore is worth $21 per ton, its market price is 70 cents per unit; and if antimony ore is $21 per ton, and the quotation is 70 cents a unit, the ore carries 30% metal.

The error is frequently made by calculating that if a unit is 20 lb., and there are 100 units in a ton, therefore the price for each unit should be multiplied by 100. This is entirely incorrect, and as explained before, the market price is based on a fixed percentage of metal. The following are a few examples:

ORE	Units or per cent of mineral in ore	Market price per unit	Fixed base, per cent.	Value per ton
Chromite	40	$0.60	40	$24.00
Fluorspar	70	0.30	80	21.00
Manganese	35	0.50	40	17.50
Pyrite	35	0.15	40	14.00
Tungsten	55	9.00	60	495.00
Iron	45	0.12	55	5.40

On other ores a flat rate is made, such as 80 cents a pound for molybdenite carrying 85%; $60 per ton for lead ore basis of 80%, and $40 per ton for zinc ore basis of 60% metal. Copper ore sold to smelters is paid on the basis of so many pounds of metal per ton multiplied by the market quotation.

To find the proportional parts by weight of the elements of any substance whose chemical formula is known, the rule is: multiply together the equivalent and the exponent of each element of the compound, and the product will be the proportion by weight of that element in the substance.

Example.— Find the proportionate weight of the elements of alcohol, C_2H_6O:

Carbon $\quad C_2 =$ equivalent $12 \times$ exponent $2 = 24$
Hydrogen $H_6 =$ equivalent $1 \times$ exponent $6 = 6$
Oxygen $\quad O =$ equivalent $16 \times$ exponent $1 = 16$

Of every 46 lb. of alcohol, 6 lb. will be H; 16, O; and 24, C.

To find the proportions by volume, divide by the specific gravity.

NAMES OF CHEMICAL SUBSTANCES.

Common Names.	Chemical Names.
Aqua fortis.	Nitric acid.
Aqua regia.	Nitro-hydrochloric acid.
Blue vitriol.	Sulphate of copper.
Cream of tartar.	Bi-tartrate of potassium.
Calomel.	Chloride of mercury.
Chalk.	Carbonate of calcium.
Caustic potash.	Hydrate of potassium.
Chloroform.	Chloride of formyl.
Common salt.	Chloride of sodium.
Copperas and green vitriol.	Sulphate of iron.

Corrosive sublimate.

Dry alum.

Epsom salts.

Ethiops mineral.

Galena.

Glauber's salt.

Glucose.

Iron pyrite.

Jeweler's putty.

King's yellow.

Laughing gas.

Lime.

Lunar caustic.

Mosaic gold.

Muriate of lime.

Muriatic acid.

Niter or saltpeter.

Oil of vitriol.

Potash.

Realgar.

Red lead.

Rust of iron.

Sal-ammoniac.

Salt of tartar.

Slaked lime.

Soda

Spirits of hartshorn.

Spirits of salt.

Stucco or plaster of Paris.

Sugar of lead.

Verdigris.

Vermilion.

Vinegar.

Volatile alkali.

White precipitate.

White vitriol.

Bi-chloride of mercury.

Sulphate of aluminium and potassium.

Sulphate of magnesium.

Black sulphide of mercury.

Sulphide of lead.

Sulphate of sodium.

Grape sugar.

Bi-sulphide of iron.

Oxide of tin.

Sulphide of arsenic.

Protoxide of nitrogen.

Oxide of calcium.

Nitrate of silver.

Bi-sulphide of tin.

Chloride of calcium.

Hydrochloric acid.

Nitrate of potash.

Sulphuric acid.

Oxide of potassium.

Sulphide of arsenic.

Oxide of lead.

Oxide of iron.

Chloride of ammonium.

Carbonate of potassium.

Hydrate of calcium.

Oxide of sodium.

Ammonia.

Hydrochloric acid.

Sulphate of lime.

Acetate of lead.

Basic acetate of copper.

Sulphide of mercury.

Acetic acid (diluted).

Ammonia.

Ammoniated mercury.

Sulphate of zinc.

As remarked in the introduction to this book, prospectors should have some knowledge of the mining laws of this country, so as to guide them in their work. When an ore deposit is discovered, and a claim is located, it must be 1500 feet long and 600 feet wide, along the mineralized formation or vein. If no regular location blank forms are available, the notice may be written on a soft pine board with a hard ' lead ' pencil, which will stand all kinds of weather. Directions in the notice should say northerly, southerly, easterly, or westerly, so as not to be too specific, in case of a desire to make deviations later on.

In the foregoing chapters no mention has been made of what a prospector should do when an orebody has been discovered, as it is assumed that he understands the various steps in development. Unless the topography of the country prevents it, the first work is to sink a small shaft on the ore-shoot, then driving on it at a depth of 50 or 100 feet. If the pay-ore persists, and the ground is hilly, a tunnel may be driven. Construction of a windlass for the shaft is simple. It is a useful contrivance, but has an economic limit at 100 feet. There are several good makes of small hoists driven by oil engines. A whip or whim may also be used. If it is desired to treat gold ore on the claim, a 2-stamp mill with amalgamating table will suffice for free-milling material. There are several good makes of these little mills. For other ores, even gold, a small ball-mill and concentrator, driven by a distillate engine, will be found entirely satisfactory.

GLOSSARY OF TERMS

Abraded. Reduced to powder.

Acicular. Needle-shaped.

Acid. An acid or silicious rock is one in which the bases are combined with silica; the opposite to basic.

Adamantine. Of diamond luster.

Adit. A nearly horizontal passage from the surface by which a mine is entered. In the United States an adit is usually called a tunnel. Adits are often the outlet for water pumped from deeper levels.

Agate. Name given to certain silicious minerals; a variegated chalcedony.

Aggregation. A coherent group.

Alkali. The opposite of an acid. It turns litmus paper or solution blue, and forms salts with acids. Potash and soda are the common alkalis.

Alloy. A mechanical compound of two or more metals fused together.

Alluvium. An earthy or gravel deposit made by running streams, especially in times of flood.

Amalgamation. The production of an amalgam or alloy of mercury; also the process in which gold and silver are extracted from pulverized ores by producing an amalgam from which the mercury is afterwards expelled by distillation.

Amorphous. Without any crystallization or definite form.

Amygdaloids. Small almond-shaped vesicular cavities in certain igneous rocks, partly or entirely filled with other minerals, such as calcite and agate.

Analysis (in chemistry). An examination of a substance to find out the nature of the component parts and their quantities; the former is called qualitative and the latter quantitative analysis.

Anemometer. An instrument for measuring the rapidity of an air-current.

Anticlinal. The line of a crest, above or under ground, on the two sides of which the strata dip in opposite directions. The converse of synclinal.

Apex. The end or edge of a vein nearest the surface. The law of the apex has been responsible for many millions of dollars being spent in lawsuits in the United States.

Aqua regia. A mixture of nitric and hydrochloric acids, in the proportion of 1 to 3.

Arborescent. Of tree-like form.

Arenaceous. Silicious or sandy (or rocks).

Argentiferous. Containing silver.

Argillaceous. Containing clay.

Arrastra. Apparatus for grinding and mixing ores by means of a heavy stone dragged around upon a circular bed. Chiefly used for ores containing free gold. At best it is a crude machine, yet much metal has been extracted thereby.

Arsenite. Compound of a metal with arsenic.

Assay. To test ores and minerals by chemical, blowpipe, or furnace examination.

Assay-ton. A weight of 29.1666 grams.

Assessment work. The work done annually on an unpatented mining claim to maintain possessory title. The value of the work is fixed at $100.

Auriferous. Containing gold.

Axe stone. A species of jade. It is a silicate of magnesia and alumina.

Back of a lode. The ground between the roof and the surface.

Bank claim. A mining claim on the bank of a stream.

Banket. Auriferous conglomerate cemented together with quartz grains. The banket of the Rand in South Africa yields 40% of the world's gold.

Bar. A vein or dike crossing a lode; also a sand or rock ridge crossing the bed of a stream.

Bar-diggings. Gold-washing claims on the bars (shallows) of a stream, and worked when the water is low, or otherwise with the aid of coffer-dams.

Barilla. Spanish term for concentrates; used largely in connection with tin in Bolivia.

Barrel amalgamation. The treatment of silver ores in wooden barrels with quicksilver, metallic iron, and water.

Base metals. The metals not classed as precious.

Bases. Compounds that are converted into salts by the action of acids.

Basic. Low in silica.

Basin. A natural depression of strata containing a coal bed or other stratified deposit; also the deposit itself. A syncline.

Battery. A set of stamps in a stamp-mill comprising the number that fall in one mortar, usually five; also a bulkhead of timber. A modern 2-stamp prospector's mill is an admirable machine to have on a small mine.

Battery amalgamation. Amalgamation by means of mercury placed in the mortar.

Bed. A seam or deposit of mineral, later in origin than the rock below, and older than the rock above; that is to say, a regular member of the series of formation, and not an intrusion.

Bedded vein. A lode occupying the position of a bed, that is, parallel with the stratification of the enclosing rocks.

Bedrock. The solid rock underlying alluvial and other surface formations.

Bed-way. An appearance of stratification, or parallel marking, in granite.

Belly. A swelling mass of ore in a lode.

Black band. A variety of carbonate of iron.

Black flux. A mixture of charcoal and potassium carbonate.

Black jack. Zinc blende, sphalerite.

Black tin. Tin ore ready dressed for smelting, usually containing over 70% of metal.

Blanch. Lead ore mixed with other minerals.

Blanched copper. An alloy of copper and arsenic.

Blende. Sulphide of zinc, from 'blenden' to dazzle.

Blind level. A level not yet connected with other workings.

Blind lode. One that does not show surface outcrops.

Blind shaft. An underground shaft; one that does not reach the surface.

Blossom. The oxidized or decomposed outcrop of a vein or coal bed.

Blow-out. (1) A large outcrop beneath which the vein is smaller. (2) A shot or blast is said to blow out when it goes off like a gun, and does not shatter the rock.

Blue-john. Fluorspar.

Blue lead. The bluish auriferous gravel and cement deposit found in the ancient river-channels of California.

Bluestone. Copper sulphate.

Bluff. A high bank or hill with a precipitous front.

Bonanza. A body of rich ore, generally used in speaking of gold and silver.

Booming. The accumulation and sudden discharge of a quantity of water, as in placer-mining, where water is scarce. See 'Hushing.'

Bort. Opaque black diamond.

Botryoidal. Like a bunch of grapes.

Boulder. A fragment of rock brought by natural means from a distance, and usually large and rounded in shape.

Box canyon. A canyon closed at one end.

Brasque. A lining for crucibles; generally a compound of clay, etc., with charcoal dust.

Breast. The face of a working.

Breccia. A conglomerate in which the fragments are angular.

Buddle. A conical table, or stationary or revolving platform upon which ore is concentrated by means of running water.

Bulkhead. An underground dam or partition of wood or concrete built to keep back water or bad ground.

Bullion. Gold and silver mixed, or these with any of the base metals.

Buried rivers. River beds that have been buried below streams of basalt or alluvial drifts, such as in California and in Victoria, Australia.

Butte. A hill.

Button. The globule of metal remaining in a crucible or cupel at the end of fusion.

Cage. A frame with one or more platforms used in hoisting men and ore cars in a vertical shaft.

Calcareous. Containing carbonate of lime; or a limey substance.

Calcination. Roasting to drive off sulphur or gases; frequently called burning.

Calcine. To expose to heat with or without oxidation.

Calcite. Carbonate of lime.

Cañon. A valley, usually precipitous; a gorge.

Cap or **cap-rock.** Rock above an ore deposit; often called overburden.

Carat. Weight, 3.2 grains, used for diamonds and precious stones. With goldsmiths and assayers the term carat is applied to the proportions of gold in an alloy; 24 carats represents fine gold. Thus 18-carat gold signifies that 18 out of 24 parts are pure gold, the remainder some other metal. The fineness of gold is obtained by dividing 24 into the carat value: thus, 24 into $18 = 750$ fine.

Carbonaceous. Containing carbon not oxidized.

Carbonates. The common term in the West for ores containing a considerable proportion of carbonate of lead.

Carboniferous. Rocks of the coal age of geologic eras.

Carbonization. Conversion to carbon.

Casing. Clayey material found between a vein and its wall.

Cawk. Sulphate of baryta (heavy spar).

Cement. Gravel firmly held in a silicious matrix, or the matrix itself.

Chert. Flint or chalcedony. A silicious stone often found in limestone, such as in southwestern Missouri and Oklahoma.

Chimney. An orebody of pipe shape in an approximately vertical position.

Choke damp. Carbonic acid gas.

Chlorides. A common term for ores containing chloride of silver.

Chloridize. To convert into chloride. Applied to the roasting of silver ores with salt, preparatory to amalgamation.

Chute. A channel or shaft underground, or an inclined trough above ground, through which ore falls or is dropped by gravity from a higher to a lower level.

Claim. The portion of mining ground held under the Federal and local laws by one claimant or association, by virtue of one location and record. A full claim is 600 feet wide and 1500 feet long, and contains 20 acres.

Clay slate. A slate formed by the induration of clay.

Cleavage. The property of a mineral of splitting more easily in some directions than in others.

Cleavage planes. The planes along which cleavage takes place.

Clinometer. An apparatus for measuring vertical angles, particularly dips.

Collar. The top of a shaft. Large shafts are now concreted.

Color. A particle of gold found in the prospector's pan.

Concentration. The removal by mechanical means of the metallic part of an ore from the lighter and less valuable portions. The ore must be crushed fine, then subjected to separation by dry or wet means — gravity, magnetic, water, or oil flotation.

Conchoidal. Name given to a certain kind of fracture resembling a bivalve shell.

Concretion. A nodule formed by the aggregation of mineral matter from without round some center.

Conglomerate. A rock consisting of fragments of other rocks, usually rounded, cemented together. Important examples are those at Cobalt, Ontario; the Michigan copper country; and the Rand, South Africa.

Contact. The plane between two adjacent bodies of dissimilar rock. A contact-vein is a vein, and a contact-bed is a bed, lying, the former more or less closely, the latter absolutely, along a contact.

Contortion. Crumpling and twisting.

Coprolites. Phosphate of lime; petrified excrements of animals.

Costeen. A trench across a lode on the surface.

Counter. A cross vein.

Country, or country rock. The rock traversed by or adjacent to an ore deposit.

Course of a lode. Its direction or strike.

Cradle. See rocker.

Crate dam. A dam built of crates filled with stone.

Crater. The cup-like cavity at the summit of a volcano.

Cretaceous. A geologic period; chalky.

Crevice. A shallow fissure in the bedrock under a gold placer, in which .small but highly concentrated deposits of gold are found; also the fissure containing a vein.

Crevasse. A deep fissure in ice or in a glacier.

Cribbing. Close timbering, as the lining of a shaft, or in a drift.

Cropping-out. The rising of layers of rock to the surface.

Cross-course. An intersecting vein.

Crosscut. A level driven across the course of a vein.

Cross-vein. An intersecting vein.

Cupriferous. Containing copper.

Cyanidation. Dissolving gold or silver by cyanide of potassium or sodium and precipitating it on zinc or aluminum.

Datum level. The point (usually sea-level) from which altitudes are measured in survey.

Dead-roasting. Roasting carried to the farthest practicable degree in the expulsion of sulphur.

Dead-work. Work that is not directly productive, though it may be necessary for exploration and future production.

Debris. The fragments resulting from shattering and disintegration.

Decomposed. Alteration by air and water.

Decrepitate. To crackle and fly to pieces when heated.

Deep leads. Alluvial deposits of gold or tinstone buried below a considerable thickness of soil or rock, the latter usually of volcanic origin.

Delta. The alluvial land at the mouth of a river; usually bounded by two branches of the river, so as to be of a more or less triangular form.

Dendritic. Like branches of trees.

Denudation. Rock laid bare by water or other agency.

De-oxidation. The removal of oxygen.

De-silverization. The process of separating silver from its alloys.

De-sulphurization. The removal of sulphur from sulphide ores.

Detritus. Accumulations from the disintegration of exposed rock surfaces.

Development. Work done in opening an ore deposit, not mining or exploration.

Diabase. A dark trap rock composed of crystals of feldspar, with ferro-magnesian minerals.

Dialling. Surveying a mine by means of a dial.

Diggings. Applicable to all mineral deposits and mining camps, but used in the United States when applied to placer-mining only.

Dike. A vein of igneous rock.

Diluvium. Sand, gravel, clay, etc., in surficial deposits.

Dip. The inclination of a vein or stratum below the horizontal.

Discovering. The first finding of the mineral deposit in place upon a mining claim. A discovery is necessary before the location can be held by a valid title. The opening in which it is made is called discovery shaft, discovery tunnel, etc.

Disintegration. The breaking asunder of solid matter due to chemical or physical forces.

Dislocation. The displacement of rocks on either side of a crack.

Disseminated ore. That which carries fine particles of sulphide minerals scattered through rock or gangue. The great porphyry copper mines are disseminated deposits.

Ditch. An artificial water-course, flume, or canal to convey water for mining.

Domes. Strata that dip away in every direction; an anticline.

Drift. A horizontal passage underground, following a vein; also unstratified diluvium.

Druse. A crystallized crust lining the sides of a cavity in rock or ore.

Dry ores. Silver ores that do not contain lead.

Ductile. Any metal that can be drawn into wire is ductile, such as copper.

Efflorescence. An incrustation of powder or threads, due to the loss of the water of crystallization.

Element. One of the fundamental substances of the earth; something that can not be decomposed under ordinary circumstances. The metals are all elements.

Elutriation. Purification by washing and pouring off the lighter matter suspended in water, leaving the heavier portions behind.

Entry. An adit.

Erosion. The act or operation of wearing away by water, wind, or glacial action.

Eruptive. Igneous rocks brought to the earth's surface or outer crust by volcanic action.

Excrescence. Grown out from something else.

Exfoliate. To peel off in leaves from the outside.

Exploitation. The working of a property as distinguished from exploration or prospecting.

Face. In any adit, tunnel, or slope, the end at which work is progressing or was last done.

False bottom. In alluvial mining a stratum on which auriferous beds lie, but which has other bottoms below it.

Fathom. 6 feet.

Fault. A dislocation of the strata or vein. Many veins are lost or lose their value through faults.

Feather ore. A sulphide of lead and antimony.

Feeder. A small vein adjoining a larger vein.

Feldspathic. Containing feldspar as the principal ingredient.

Ferruginous. Containing iron.

Fire-damp. Light carburetted hydrogen gas.

Fissile. Splitting easily into plates.

Fissure-vein. A fissure in the earth's crust filled with mineral matter.

Flexible. Capable of being bent without elasticity.

Flint. A massive impure variety of silica; chert or chalcedony.

Float-copper. Fine scales of metallic copper that do not readily settle in water.

Float gold. Fine particles of gold that do not readily settle in water, and hence are liable to be lost in ordinary stamp-milling, but may be recovered by cyanidation afterwards.

Float ore. Water-worn particles of ore; particles of vein-material found on the surface, away from the vein outcrop. By following float ore, deposits are traced.

Flocculent. Cloudy, resembling lumps of wool.

Floor. The rock underlying a stratified or nearly horizontal deposit, also a horizontal flat orebody.

Flotation. A process of concentrating ores by mixing the pulp (ore and water) with acid or oil, and agitating the whole by air or paddles to make a scum or froth, which takes up the metallic minerals and floats them off, leaving the waste behind.

Flume. A wooden conduit bringing water to a mine or mill.

Flux. A salt or other mineral added in assaying or smelting to assist fusion by forming more fusible compounds.

Foliated. Arranged in leaf-like laminæ (such as mica-schist).

Foot-wall. The wall under the vein.

Forfeiture. The loss of possessory title to a mine by failure to comply with the laws prescribing the quantity of assessment work, or by actual abandonment.

Formation. The series of rocks belonging to an age, period or epoch, such as the Silurian formation. The term is used in mines to describe any particular rock in which a vein may exist.

Fossil. Term applied to express the animal or vegetal remains found in rocks.

Free. Native, uncombined with other substances, as free gold or silver.

Free-milling. Applied to ores that contain free gold or silver, and can be reduced by crushing and amalgamation, without roasting or other chemical treatment.

Friable. Any substance that may be broken easily is friable.

Fuller's earth. An unctuous clay, usually of a greenish-gray tint, compact yet friable. Used for filtering oils and greases.

Fuse. In blasting, the fire is conveyed to the blasting agent by means of a prepared tape or cord called the fuse.

Gad. A steel wedge.

Gallery. A level or drift.

Gangue. The waste rock mineral associated with the ore in a vein.

Gash. Applied to a vein wide above, narrow below, and terminating in depth within the formation it traverses.

Geode. A cavity, studded around with crystals or mineral matter, or a rounded stone containing such cavity. See vug.

Geology. The science of the formation of the earth and the rocks therein.

Geysers. Intermittent boiling springs.

Glacier. A body of ice that descends from high to low ground. They move very slowly, sometimes only a few inches a year.

Glance. Literally, shining. Name applied to certain sulphides — copper glance is chalcocite; silver glance is argentite.

Globule. A small substance of spherical shape.

Glory-hole. A large open pit from which ore is extracted by allowing it to fall to a lower level in the mine, whence it is hauled to the surface.

Gneiss (nice). A banded coarse-grained rock often formed of the same minerals as granite, but of irregular arrangement.

Gopher or **gopher-drift.** An irregular prospecting drift, following or seeking the ore without regard to maintenance of a regular grade or section.

Gossan. Hydrated oxide of iron, usually found at the decomposed outcrop of a mineral vein.

Gouge. A thin seam of clay separating ore, or ore and rock.

Gravel mine. In the United States, an accumulation of auriferous gravel.

Granulated. Grain-like.

Greenstone. An altered basic porphyry, such as andesite or diorite.

Grit. A variety of sandstone of coarse texture.

Gubbin. A kind of ironstone.

Gulch. A ravine.

Gullet. An opening in the strata.

Gut. To rob a mine of its rich ore.

Hade. See underlay.

Hanging-side or **Hanging-wall**, or **Hanger.** The wall or side over the vein.

Heading. The face or breast of a drift or other working in a mine.

Heading side. The under side of a lode.

Head-frame. The structure erected over a shaft, by means of which hoisting may be performed.

Heave. An apparent lateral displacement of a lode produced by a fault.

Hog back. A sharp anticlinal, decreasing in height at both ends until it runs out; also a ridge produced by highly tilted strata.

Homogeneous. Of the same structure throughout.

Horizon. Geologically all rock strata of the same period. Also the sky-line.

Horse. A mass of country rock enclosed in an ore deposit.

Hungry. A term applied to hard, barren vein-matter, such as white quartz.

Hushing. The discovery of veins by accumulation and sudden discharge of water, which 'washes away the surface soil and lays bare the rock. See booming.

Hydraulicking. Washing down a bank of earth or gravel by the use of giants or monitors, conveying water under high pressure.

Hydrous. Containing water in its composition.

Igneous. Resulting from volcanic action; as, lava and basalt are igneous rocks.

Impregnation. An ore-deposit consisting of country-rock impregnated with minerals.

Incline. A shaft not vertical; also a plane, not necessarily under ground.

Incrustation. A coating usually crystallized.

Indicator vein. A vein that is not metalliferous itself, but if followed, leads to ore-deposits. At Bendigo, Australia, are many indicators.

In place (in situ). Of rock, occupying, relative to surrounding masses, the position that it had when formed.

Intrusion. Forcing through.

Irestone. Hard clay slate; hornstone; hornblende.

Iridescent. Showing rainbow colors.

Jack. Black jack, sphalerite, or zinc blende.

Jigging. Separating ores according to specific gravity with a sieve agitated up and down in water. The apparatus is called a jig.

Jinny-road. A gravity plane underground.

Jump. To take possession of a mining claim alleged to have been forfeited or abandoned; also, a dislocation of a vein.

Kibble. An iron bucket for raising ore in a shaft.

Kicker. Ground left in first cutting a vein, for support of its sides.

Kindly. When a rock has good mineralization it is termed kindly.

King's yellow. Sulphide of arsenic.

Knits or **knots.** Small particles of ore.

Lagging. Small timber placed above sets in drifts, or around those in shafts.

Lagoon. A marsh, shallow pond or lake.

Lamellar. In thin sheets.

Lamina. A thin plate or scale.

Lava. Rock formed by the consolidation of liquid matter which has flowed from a volcano.

Leaching. The separation of a soluble from an insoluble material by means of a solvent; for instance, dissolving gold from quartz by cyanide, or dissolving copper by sulphuric acid.

Leads. The auriferous portions of alluvial deposits marking the former courses of streams.

Leath. Applied to the soft part of a vein.

Ledge. A deposit of oil-shale is the only true ledge. A vein is frequently incorrectly called a ledge.

Lenticular. Lens-like.

Level. A horizontal passage or drift into or in a mine.

Litharge. Protoxide of lead.

Loadstone. An iron ore consisting of protoxide and peroxide of iron; magnetite.

Locate. To establish the right to a mining claim. This word is frequently used incorrectly in place of 'find' or 'discover,' and 'location,' is incorrectly used for 'position' or 'situation.'

Lode. A metalliferous orebody that has no well-defined walls and may not be homogeneous as distinguished from a vein, which see.

Long tom. A gold-washing cradle.

Magna. Groundwork of igneous rocks.

Mainway. A gangway or principal passage.

Marl. Clay containing carbonate of lime.

Mass-copper. Native copper occurring in large lumps.

Massicot. See litharge.

Matrix. The rock or earthy material containing a mineral or metallic ore; the gangue.

Matte. A furnace product carrying copper and iron sulphides, usually 30 to 40% copper.

Measures. Strata of coal, or the formation containing coal beds.

Mesh. Openings in a screen.

Metalliferous. Metal-bearing.

Metallurgy. The science of ore treatment.

Metamorphic. Changed in form and structure.

Metasomatic. Replacement, particle by particle, so that the original structure is often preserved.

Mine. In general, any excavation for minerals. More strictly, subterranean workings, as distinguished from quarries, placer and hydraulic mines, and surfaces or open works.

Mineralized. Charged or impregnated with metalliferous mineral.

Mineral-right. The ownership of the minerals under a given surface, with the right to enter thereon, mine and remove them. It may be separated from the surface ownership, but, if not so separated by distinct conveyance, the latter includes it.

Mine-rent. The rent or royalty paid to the owner of a mineral right by the operator of the mine.

Miner's inch. A local unit for the measurement of water supplied to hydraulic miners. It is the amount of water flowing under a certain head through one square inch of the total section of a certain opening for a certain number of hours daily.

Minium. Proto-sesqui-oxide of lead.

Mock ore. A false kind of mineral, such as mock opal.

Moil. A piece of steel, like a drill, sharpened to a point. It is used largely in sampling ore.

Monkey drift. A small prospecting drift.

Monoclinal. Applied to any limited portion of the earth's crust throughout which the strata dip in the same direction.

Monzonite. A granitic rock, a type between diorite and granite.

Mountain blue. Blue copper ore.

Muffle. A semi-cylindrical or long-arched oven, usually small and made of fireclay, used in assaying.

Mundic. Iron pyrite, called so in Cornwall and Australia. White mundic is mispickel.

Nacreous. Resembling mother-of-pearl.

Native. Occurring in nature; not artificially formed; usually applied to the metals.

Nickeliferous. Containing nickel.

Noble metals. The metals which have so little affinity for oxygen that their oxides are reduced by the mere application of heat without a reagent; in other words, the metals least liable to oxidation under ordinary conditions. The list includes gold, silver, mercury, and the platinum group.

Nodule. A small concretionary rounded mass of mineral matter.

Non-conformable. Rock strata not associated originally in the position now occupied.

Nugget. A lump of native metal, especially of a precious metal.

Nucleus. A body about which anything is collected.

Open-cut. A surface working, open to daylight.

Ore. Rock that carries enough mineral to be profitably exploited is ore.

Ore vertical. An ore deposit peculiar to the Black Hills of South Dakota.

Ore-shoot. That part of a vein which contains pay-ore.

Organic compounds. Those containing carbon, generally derived from animals or plants.

Outcrop. The portion of a vein or stratum emerging at the surface, or appearing immediately under the soil and surface debris.

Output. The product of a mine.

Overburden. The cap-rock or waste that usually covers an ore deposit.

Oxidation. A chemical union with oxygen.

Oxide. The combination of a metal with oxygen.

Pack walls. Walls built of loose material in mines to support the roof.

Panning. Washing earth or crushed rock in a pan, by agitation with water, to obtain the particles of greatest specific gravity it contains; chiefly practised for gold, also for quicksilver, diamonds, gems and other minerals.

Parting. The separation of two metals in an alloy, especially the separation of gold and silver by means of nitric or sulphuric acid.

Patent. A deed conveyed by the Government to a miner that has complied with all the regulations concerning his claim.

Pay-streak. The zone in a vein that carries the profitable or pay-ore.

Peroxide. An oxide containing more oxygen than some other oxide of the same element.

Peter-out. To fail gradually in size or quality, to pinch out.

Petrified. Changed to stone.

Petrology. Study of rocks by means of thin sections and the microscope.

Phosphates. Phosphoric acid combinations.

Pinch. To contract in width.

Pipe or **pipe-vein.** An orebody of elongated form. A famous pipe in Western Australia yielded over $30,000,000 of gold.

Piping. Washing gold deposits by means of a hose.

Placer. A deposit of valuable mineral, found in particles in alluvium or diluvium, or beds of streams, etc.

Plastic. Easily moulded.

Plat. The map of a survey in horizontal projection. In Australia it also means a shaft station.

Plumbago. Graphite or black lead.

Plumb-bob. A weight suspended by a string to determine vertical lines.

Plush Copper. A fibrous red copper ore.

Pocket. A small body of ore, or a cavity cut in rock to hold ore.

Porphyritic. Of the nature of porphyry, carrying isolated crystals in a ground mass.

Potstone. Compact steatite.

Precipitate. When a metal has been dissolved, and it is desired to throw it down in another form, this is called precipitation.

Prill. A good-sized piece of pure ore, or metal scattered throughout a slag.

Prisms. Solids whose bases are plane figures, and whose sides are parallelograms.

Prospect. A new property that has not been developed enough to be called a mine.

Prospector. One who searches for minerals.

Pryan. Ore in small pebbles mixed with clay.

Pudding-stone. A conglomerate in which the pebbles are rounded.

Putty powder. Crude oxide of tin.

Quarry. An open pit from which stone or ore is mined.

Quartz. Crystalline silica; also, any hard gold or silver ore, as distinguished from gravel or earth, hence quartz-mining as distinguished from hydraulic mining.

Quartzite. An altered crystalline sandstone.

Quartzose. Containing quartz as a principal ingredient.

Quicksand. Sand that is, or becomes upon the access of water 'quick,' that is, shifting, easily movable or semi-liquid.

Radiating. Diverging from a center.

Raise. A shaft that is driven upwards from a level; sometimes called a rise.

Rake. Pitch of an ore-shoot on the plane of the vein.

Range. A mineral-bearing belt of rocks, such as in the Lake Superior iron region, also Wisconsin zinc fields.

Ravine. A deep narrow valley.

Reduce. To deprive of oxygen; also, in general, to treat metallurgically for the production of metal.

Reef. An incorrect term for lode or vein, used in Australia and Africa.

Refractory. Resisting the action of heat and chemical agents.

Reniform. Kidney-like.

Reticulated veins. Veins traversing rocks in all directions.

Reverse faults. Faults due to thrust; the hanging-wall side of the fault being forced upwards on the foot-wall.

Riffle. A groove or interstice, or a cleat or block, so placed as to produce the same effect, in the bottom of a sluice, to catch free gold.

Rim-rock. The bedrock rising to form the boundary of a placer or gravel deposit.

Roasting. Calcination, usually with oxidation.

Rocker. A short trough in which auriferous sands are agitated by oscillation in water, to collect the gold contained.

Roof. The strata immediately over a mine opening.

Royalty. The money paid a mine-owner by a lessee. There are several systems of payment.

Rusty gold. Free gold that does not easily amalgamate, the particles being coated, as is supposed, with oxide of iron. Grinding usually rubs off this surficial coating.

Saddle. An anticlinal in a bed or flat vein; a depression or U-shaped fold, the reverse of an arch or anticline.

Sal-ammoniac. Chloride of ammonia.

Saline. A salt spring or well; salt works.

Salvage or Selvedge. A layer of clay or decomposed rock along a vein-wall.

Sampling. Cutting a representative part of a vein, and mixing the ore so that a portion taken from the whole may fairly represent the section.

Schist. Crystalline rock, usually micaceous.

Schorl. Black tourmaline.

Seam. A stratum or bed of coal or other mineral.

Sectile. Easily cut.

Secondary enrichment. An enrichment of a vein or orebody by material of later origin, often derived from the oxidation of decomposed overlying ore masses. Theory first propounded by Weed in 1899.

Sediment. A deposit formed by water.

Segregation. A mineral deposit formed by concentration from the adjacent rock.

Shaft. A pit sunk from the surface.

Shake. A cavern, usually in limestone; also a crack in a block of stone.

Shale. Consolidated clay.

Shear-zone. A belt in which rocks are crushed by many parallel fissures.

Shingle. Clean gravel.

Shoot. See ore-shoot.

Side-basset. A transverse direction to the line of dip in strata.

Silicates. Compounds of silica or silicic acid with a base.

Silicious. Consisting of or containing silica or quartz.

Slag. The vitreous mass separated from the molten metals in smelting ores.

Slate. Indurated clays, sometimes metamorphosed.

Slickensides. Polished and sometimes striated surfaces on the walls of a vein, or on interior joints of the vein-material or of rock masses.

Slide. A fault or cross-course.

Sline. Natural transverse cleavage of rock.

Slip. A vertical dislocation of rocks.

Slope. An inclined opening to a mine.

Sluicing. Washing auriferous earth.

Sollar. A platform or ladder landing in a shaft.

Spall. To break ore. Pieces or ore thus broken are called spalls.

Speiss (spice). Impure metallic arsenides, principally of iron, produced in copper and lead smelting. Cobalt and nickel are found concentrated in the speiss obtained from ores containing these metals.

Spelter. Trade name for zinc.

Spoon. An instrument made of an ox or buffalo horn, in which earth or pulp may be carefully tested by washing to detect gold, amalgam, etc.

Spur. A branch leaving a vein, but not returning to it.

Stalactites. Icicle-like incrustations hanging down from the roof of caves.

Stalagmites. Similar to stalactites, but formed on the floor of the caves by the deposition of solid matter held in solution by dropping water.

Stannary. A tin mine, or tin works.

Step-vein. A vein alternately cutting through the strata of country-rock and running parallel with them.

Stockwerk. An ore deposit consisting of many small stringers that could not be mined separately but the mass is rich enough to be extracted as a whole.

Stope. To remove ore from above a drift.

Stratum. A bed or layer.

Streak. The powder of a mineral, or the mark which the latter makes when rubbed upon or by a harder substance.

Striated. Marked with parallel grooves.

Strike. The direction of a horizontal line drawn in the middle plane of a vein or stratum not horizontal.

String. A small vein.

Strip. To remove from a quarry, or open working, the overlying earth and disintegrated or barren surface rock.

Stull. A platform laid on timbers, braced across a working from side to side, to support workmen or to carry ore or waste.

Sublimation. The volatilization and condensation of a solid substance without fusion.

Sub-metallic. Of imperfect metallic luster.

Subsidence. Sinking down.

Sub-transparent. Of imperfect transparency.

Sulphate. A salt containing sulphuric acid.

Sulphide. A combination of a metal with sulphur.

Sulphurets. In miners' phrase, the undecomposed metallic ores, usually sulphides. Chiefly applied to auriferous pyrite.

Surficial. Pertaining to the surface of the ground.

Synclinal. The axis of a depression of the strata; also the depression itself. Opposed to anticlinal, which is the axis of an elevation.

Tailing. The lighter and sandy portion of an ore from a concentrator; residue. Tailings from different machines are often re-treated.

Tail-race. The channels in which tailing suspended in water, are conducted away.

Texture. A rock structure; fine or coarse grained.

Thermal. Hot, such as, thermal springs.

Throw. A dislocation or fault of a vein or stratum, which has been thrown up or down by the movement.

Tinstone. Ore containing small grains of oxide of tin.

Toadstone. A kind of trap-rock.

Toughening. Refining, as of copper or gold.

Translucent. Allowing light to pass through, yet not transparent.

Trap. In miners' parlance, any dark igneous, or apparently igneous, or volcanic rock.

Trend. The course of a vein.

Tuff or **Tufa.** A soft sandstone or calcareous deposit.

Tunnel. A nearly horizontal underground passage, open at both ends to daylight. Mines in hilly country are generally opened by tunnels.

Turn. A pit sunk in a drift.

Underlay or **Underlie.** The departure of a vein or stratum from the vertical, usually measured in horizontal feet per fathom of inclined depth.

Unstratified. Not arranged in strata.

Upcast. A shaft having an upward air current.

Vanning. Washing tin-stuff by means of a shovel.

Vein. A metalliferous orebody of homogeneous formation lying between well-defined walls. See Lode.

Vein-stuff. Ore associated with gangue.

Vermilion. Mercury sulphide.

Vertical. An ore deposit peculiar to the Black Hills of South Dakota.

Vitreous. Glassy.

Volatile. Capable of easily passing off as vapor.

Vug, Vugg, or **Vugh.** A cavity in the rock, usually lined with a crystalline incrustation. See geode.

Walls. The boundaries of a lode, the upper one being the hanging, the lower the foot-wall.

Wash dirt. Auriferous gravel, sand, clay, etc.

Water-level. That point above which water in a mine does not rise. Below this point the unaltered sulphide ores commence.

Weathering. Changing under the effect of continual exposure to atmospheric agencies.

Whim. A machine for hoisting by means of a vertical drum, revolved by horse or steam power.

White-damp. A poisonous gas sometimes encountered in coal mines.

Wild lead. Zinc blende.

Windlass. A hand worked hoist consisting of a rope coiled on an 8-inch wooden drum turned by crank handles. It is the first machine used in sinking a shaft.

Wing dams. Dams built from the side of a river with the object of deflecting it from its course.

Winze. An interior shaft, usually connecting two levels, and often put down to prospect a vein at depth instead of sinking the main shaft.

Zinciferous. Zinc-bearing.

Zinc-scum. The zinc-silver alloy skimmed from the surface of the bath in the process of de-silverization of lead by zinc.

Zinc-white. Oxide of zinc.

CHEMICAL ELEMENTS, THEIR SYMBOLS AND ATOMIC WEIGHTS.*

Name of Element	Symbol	Atomic Weight	Name of Element	Symbol	Atomic Weight
Aluminum	Al	27.1	Neodymium	Nd	144.3
Antimony	Sb	120.2	Neon	Ne	20.2
Argon	A	39.88	Nickel	Ni	58.68
Arsenic	As	74.96	Niton (Radium	Nt	222.4
Barium	Ba	137.37	emanation)		
Bismuth	Bi	208.0	Nitrogen	N	14.01
Boron	B	11.0	Osmium	Os	190.9
Bromine	Br	79.92	Oxygen	O	16.00
Cadmium	Cd	112.40	Palladium	Pd	106.7
Caesium	Cs	132.81	Phosphorus	P	31.04
Calcium	Ca	40.07	Platinum	Pt	195.2
Carbon	C	12.00	Potassium (Ka-	K	39.10
Cerium	Ce	140.25	lium)		
Chlorine	Cl	35.46	Praseodymium	Pr	140.6
Chromium	Cr	52.0	Radium	Ra	226.4
Cobalt	Co	58.97	Rhodium	Rh	102.9
Columbium	Cb	93.5	Rubidium	Rb	85.45
Copper	Cu	63.57	Ruthenium	Ru	101.7
(Cuprum)			Samarium	Sa	150.4
Dysprosium	Dy	162.5	Scandium	Sc	44.1
Erbium	Er	167.7	Selenium	Se	79.2
Europium	Eu	152.0	Silicon	Si	28.3
Fluorine	F	19.0	Silver	Ag	107.88
Gadolinium	Gd	157.3	(Argentum)		
Gallium	Ga	69.9	Sodium	Na	23.00
Germanium	Ge	72.5	(Natrium)		
Glucinum	Gl	9.1	Strontium	Sr	87.63
Gold (Aurum)	Au	197.2	Sulphur	S	32.07
Helium	He	3.99	Tantalum	Ta	181.5
Holmium	Ho	163.5	Tellurium	Te	127.5
Hydrogen	H	1.008	Terbium	Th	159.2
Indium	In	114.8	Thallium	Tl	204.0
Iodine	I	126.92	Thorium	Th	232.4
Iridium	Ir	193.1	Thulium	Tm	168.5
Iron	Fe	55.84	Tin (Stannum)	Sn	119.0
Krypton	Kr	82.92	Titanium	Ti	48.1
Lanthanum	La	139.0	Tungsten	W	184.0
Lead	Pb	207.10	(Wolfram)		
(Plumbum)			Uranium	U	238.5
Lithium	Li	6.94	Vanadium	V	51.0
Lutecium	Lu	174.0	Xenon	Xe	130.2
Magnesium	Mg	24.32	Ytterbium (Neo-	Yb	172.0
Manganese	Mn	54.93	ytterbium)		
Mercury	Hg	200.6	Yttrium	Yt	89.0
(Hydragarum)			Zinc	Zn	65.37
Molybdenum	Mo	96.0	Zirconium	Zr	90.6

*Journal American Chemical Society, Vol. 27, No. 8, Aug., 1914.

The figures indicating the proportions by weight in which the elements unite with one another are called the combining or atomic weights, because they represent the relative weights of the atoms of the different elements. Since hydrogen is the lightest element, it is taken as the standard, and its combining or atomic weight = 1.

INDEX

Calamine, 190
Calaverite, 118
Calcite silver-bearing veins at
Batopilas, Mexico, and Cobalt, Ontario, 161
Calculating amount of gold in
ore from certain tests, 123,
135
Calculating averages of samples, 42
California, auriferous belt in,
138
borax in, 268
chronite in, 204
dredging in, 132
gold pocket mining, 148
magnesite in, 282
manganese in, 210
molybdenite in, 212
Mother Lode gold deposits,
148
Penn mine, 178
potash in, 286
quicksilver deposits in, 231
scheelite in, 215
Shasta copper belt, 178
tourmaline in, 306
use of rocker, 123
Canada, peat in, 273
Cannel coal, 272
Carbonate of copper, 171
of lead, 184
of zinc, 189
Carbonates, 7
and hydrochloric acid, 86
Carbon in diamonds, 294
in graphite, 278
Carnelian, 310
Carnotite, 218
Casing or gouge, what is, 39
Cassiterite, 216, 223
Central City, Colorado, gold
deposits of, 152
Cerargyrite, 155
Cerro de Pasco, Peru, copper
deposit, 178
Cerussite, 184
Ceylon, graphite in, 277

Chalcedony, 275, 309
Chalcocite, 168
Chalcopyrite, 169
silver in, 158
Chalk, 280
Chemical changes in minerals
during tests, 54
elements, symbols, and
weights, 350
substances, names of, 324
Chert, 275
ferruginous, 202
manganiferous, 210
Chile, borax in, 268
niter in, 284
porphyry copper deposit, 168
China, antimony deposits of,
193
clay, 270
salt in, 287
Chloanthite, 238
Chlorite, 14
Chromite, characteristics, 203
field tests, 205
in beach sands, 204
Chrysocolla, 170
Chrysoprase, 310
Chuquicamata, Chile, copper
deposit, 168
Cinnabar, 230
Clays, 270
Cleanliness in sampling, 42, 102
Cleavage of minerals, 19
Climax molybdenite mine,
Colorado, 212
Coal, 271
Cobalt minerals, 238
Cobalt, Ontario, cobalt at, 238
geology of, 3
silver deposits of, 161
Cobaltite, 238
Coeur d'Alene, Idaho, lead de-
posits, 187
Colemanite, 269
Colombia, gold and platinum
in, proportions of, 114
gold placers in, 147
platinum output of, 163

Real del Monte mine, Mexico, silver deposit, 159
Realgar, 265
Red and brown hematites, 198
Reducing flame, 57
Reef, what is a, 39
Reefs of the Rand, 148
Retorting amalgam, 134
Rhodesia, chromite in, 204
  platinum in, 163
Rhodochrosite, 208
Rhyolite as rock for silver veins, 159, 160
Rickardite, 120
Rico, Colorado, copper deposits, 175
Rio Tinto, Spain, pyrite and copper, 290
Rochester, Nevada, silver ores of, 156, 159, 163
Rock, definition of, 7
  -forming minerals, feldspars are, 274
  -forming minerals, micas are, 283
  salt, 286
Rocks, definition of certain, 31
  how laid down, 28
  in which oil is found, 248
Rose quartz, 311
Ruby, 300
  silver, 156
Russia, manganese in, 209
  normal output of platinum, 163
  platinum deposits of, 161
Rutile, 213

Salt, 286
  definition of, 16
Salting samples, 42, 102
Saltpeter, 284
Sampling ore, 41
Sandstone, oil in, 245
  uranium in, 218
Santa Gertrudis mine, Mexico, silver deposit of, 159

Santa Rita, New Mexico, copper deposit, 168
Sapphire, 299
Sardonyx, 312
Scales and balances, 101, 109
Scheelite, 214
Schools of mines and prospectors, 2
Scorifier, use of, 97
Scotland, oil-shale in, 259
Sedimentary rocks, 33
Serpentine, 15
  a source of platinum, 162
  and copper, 179
  and quicksilver, 232
  chromite in, 203
  diamonds in, 296
  magnesite in, 282
Shale and oil, 246
  a source of oil, 259
Shasta copper belt, California, 178
Siberia, one of great gold regions, 115
Sicily, sulphur in, 287
Siderite, 199
Silicates, 16
Silicious rocks, 7
Silver and gold ores, assay of, 103
  assay, simple, 110
  characteristics of, 152
Silver deposit of Dolly Varden mine, B. C., 159
  of Batopilas mine, Mexico, 161
Silver deposits of Cobalt, Ontario, 161
  of Idaho, 160
  of Real del Monte, Santa Gertrudis, and Dolores mines, Mexico, 159
  of Tonopah, Nevada, 159
  of Waihi mine, New Zealand, 160
Silver from silver ores, 114
  in manganese ore, 208
  minerals, 154

For references not given under this head see Glossary of terms on page 327.

www.ingramcontent.com/pod-product-compliance
Lightning Source LLC
Chambersburg PA
CBHW060318200326
41519CB00011BA/1762

* 9 7 8 1 6 1 4 7 4 0 4 8 3 *